PROCEEDINGS OF THE 30TH INTERNATIONAL GEOLOGICAL CONGRESS
VOLUME 11

STRATIGRAPHY

Proceedings of the 30th International Geological Congress

Volume   1   : Origin and History of the Earth
Volume   2   : Geosciences and Human Survival, Environment, and Natural Hazards
Volume   3   : Global Changes and Future Environment
Volume   4   : Structure of the Lithosphere and Deep Processes
Volume   5   : Contemporary Lithospheric Motion / Seismic Geology
Volume   6   : Global Tectonic Zones / Supercontinent Formation and Disposal
Volume   7   : Orogenic Belts / Geological Mapping
Volume   8   : Basin Analysis / Global Sedimentary Geology / Sedimentology
Volume   9   : Energy and Mineral Resources for the 21st Century / Geology of Mineral Deposits / Mineral Economics
Volume 10   : New Technology for Geosciences
Volume 11   : Stratigraphy
Volume 12   : Paleontology and Historical Geology
Volume 13   : Marine Geology and Palaeoceanography
Volume 14   : Structural Geology and Geomechanics
Volume 15   : Igneous Petrology
Volume 16   : Mineralogy
Volume 17   : Precambrian Geology and Metamorphic Petrology
Volume 18.A : Geology of Fossil Fuels - Oil and Gas
Volume 18.B : Geology of Fossil Fuels - Coal
Volume 19   : Geochemistry
Volume 20   : Geophysics
Volume 21   : Quaternary Geology
Volume 22   : Hydrogeology
Volume 23   : Engineering Geology
Volume 24   : Environmental Geology
Volume 25   : Mathematical Geology and Geoinformatics
Volume 26   : Comparative Planetology / Geological Education / History of Geosciences

PROCEEDINGS OF THE
30TH INTERNATIONAL GEOLOGICAL CONGRESS

BEIJING, CHINA, 4 - 14 AUGUST 1996

VOLUME 11

STRATIGRAPHY

EDITORS:

WANG NAIWEN
INSTITUTE OF GEOLOGY AND LITHOSPHERE RESEARCH CENTER, CHINESE
ACADEMY OF GEOLOGICAL SCIENCES, BEIJING, CHINA

J. REMANE
INSTITUT DE GÉOLOGIE, UNIVERSITÉ DE NEUCHÂTEL, NEUCHÂTEL, SWITZERLAND

CRC Press
Taylor & Francis Group
Boca Raton London New York

CRC Press is an imprint of the
Taylor & Francis Group, an informa business

CRC Press
Taylor & Francis Group
6000 Broken Sound Parkway NW, Suite 300
Boca Raton, FL 33487-2742

© 2017 by Taylor & Francis Group, LLC
CRC Press is an imprint of Taylor & Francis Group, an Informa business

First issued in paperback 2019

No claim to original U.S. Government works

ISBN 13: 978-0-367-44797-7 (pbk)
ISBN 13: 978-90-6764-274-3 (hbk)

Visit the Taylor & Francis Web site at
http://www.taylorandfrancis.com

and the CRC Press Web site at
http://www.crcpress.com

CONTENTS

Multidisciplinary chronostratigraphy
J. Remane   1

Some constraints on the Phanerozoic time-scale
*F.M. Gradstein, J. Ogg, F.P. Agterberg, J. Hardenbol
and P. van Veen*   11

TBO-stratotype system and Non-Smith Stratigraphy
Wang Naiwen   21

Sequencing, scaling and correlation of stratigraphic events
F.P. Agterberg and F.M. Gradstein   29

Sedimentation gaps in cyclic sequences
W. Schwarzacher   39

The application of RASC/CASC methods to quantitative
biostratigraphic correlation of Neogene in Northern South China Sea
Wang Ping and Zhou Di   45

Application of high-resolution stratigraphic correlation
approaches to fluvial reservoir
Deng Hongwen, Wang Hongliang and T.A. Cross   55

Carboniferous sequence stratigraphy and oil and gas in
Tarim Basin, Northwest China
Jiayu Gu   61

Neoproterozoic acritarch biostratigraphy of China
Yin Leiming and Yin Chongyu   67

Tremadoc trilobites from the Mungog Formation, Weongweol, Korea
D.H. Kim and D.K. Choi   75

Lower Silurian (Llandovery) rugose coral assemblage zones and their
relation with the depositional sequence of Upper Yangtze region, China
Chen Jianqiang and He Xinyi   85

Tethys - An archipelagic ocean model
Yin Hongfu   91

An integrated chronostratigraphic scheme for the Permian System
Jin Yugan   99

High frequence glacio-eustasy and carbon isotope evolution of the
Triticites Zone in South China
Liu Benpei and Li Rufeng   115

Devonian-Carboniferous Boundary in the neritic facies areas of
South China from the view-point of integrative stratigraphy
Wang Xunlian, Zhang Shilong and Xue Xiaofeng   121

Dicynodon and Late Permian Pangea
S.G. Lucas   133

Correlation of the Permian-Triassic boundary in Arctic Canada
and comparison with Meishan, China
Ch.M. Henderson and A. Baud   143

Conodont sequences and their lineages in the Permian-Triassic
boundary strata at the Meishan Section, South China
Ding Meihua, Lai Xulong and Zhang Kexin   153

Late Paleozoic deep-water facies in Guangxi, South China and
its tectonic implications
Wu Haoruo, Wang Zhongcheng and Kuang Guodun   163

Morphological change of Late Permian Radiolaria as seen in
pelagic chert sequences
K. Kuwahara   171

Clastic rocks in Triassic bedded chert of the Mino terrane,
Central Japan and the Samarka terrane, Sikhote-Alin, Russia
S. Kojima, K. Sugiyama, I.V. Kemkin, A.I. Khanchuk and S. Mizutani   181

Holarctic fossil mammals and Paleogene Series boundaries
S.G. Lucas   189

A candidate section for the Lower-Middle Pleistocene boundary
(Apennine Foredeep, South Italy)
N.P. Ciaranfi, A. D'Alessandro and M. Marino   201

The Plio-Pleistocene diatom record from ODP Site 797
of the Japan Sea
I. Koizumi and A. Ikeda   213

Middle to Late Pleistocene shallow-marine sedimentary and
faunal cycles corresponding to the orbital precession or obliquity
in the Simosa Group, Boso Peninsula, central Japan
T. Kamataki and Y. Kondo   233

Proc. 30ᵗʰ Int'l. Geol. Congr., Vol. 11 , pp. 1-9
Wang Naiwen and J Remane (Eds)
© VSP 1997

Multidisciplinary Chronostratigraphy

JÜRGEN REMANE

(Chairman of the International Commission on Stratigraphy, ICS)

Université de Neuchâtel, Institut de Géologie, 11, rue Emile-Argand, CH 2007 Neuchâtel, Switzerland

Abstract

Classical chronostratigraphy relies mainly on two methods: radiometric dating, providing numerical ages, and biochronology, based on a reference scale of relative ages, derived from the (bio)stratigraphic distribution of fossils. This standard scale of relative ages is still in general use, and one of the main tasks of ICS is to work out precise definitions of the boundaries of its subdivisions. In former times, physical methods of chronocorrelation played only a minor role. This was due to the fact that most of them were only the expression of regional and, above all, repetitive events. On the other hand, even now most of the traditional boundaries are not yet defined in a satisfactory manner, due to shortcomings of the classical methods (rarity of radioactive minerals, regional and environmental limitations of fossil species). With the development of new methods of chronocorrelation, chronostratigraphy has become a truly multidisciplinary field of research. Reversals of the Earth magnetic field, shifts in stable isotope curves, geochemical signals (anoxic events, the famous Ir spike at the K/T boundary), sequence stratigraphy, recognition of Milankovich cycles in the sedimentary record, together with the mathematical techniques of Quantitative Stratigraphy, have opened new perspectives. The point is that certain signals, especially shifts of stable isotope curves and magnetic reversals, reflect world wide events, which gives them a unique correlation potential. Magnetic reversals are practically instantaneous compared to the duration of geological time. All these signals are repetitive and cannot simply replace the classical methods. But combined with them, they give a new impetus to chronostratigraphy. Once a repetitive signal has been correctly identified, e. g. with the help of fossils, it will allow to extend and refine biostratigraphic correlations.

Keywords: chronostratigraphic boundary definitions, global events, biostratigraphy, stable isotope shifts, magnetostratigraphy

INTRODUCTION

The title of this paper may be misleading: this is not a text-book on chronostratigraphy. The intention is to give a general introduction to the Symposium 1 "Stratigraphy" of the 30ᵗʰ International Geological Congress at Beijing. This symposium was initiated by ICS (the International Commission on Stratigraphy). Individual sessions were organized by various bodies of ICS, and the diversity of subjects is quite impressive. Session 1-1, entitled (I abridge): "Multidisciplinary approaches in establishing the geochronologic scale", was meant to give an overlook over the diversity of chronostratigraphic methods. It is, so to say, the introductory session of Symposium 1, and this paper is then the introduction to the introduction. I best start with a concrete example of my own experience:

THE JURASSIC-CRETACEOUS BOUNDARY IN NORTH-EASTERN MEXICO

The field work was conducted by a team of three scientists (T. Adatte, Neuchâtel microfacies and clay minerals, W. Stinnesbeck, Linares, Mexico, ammonites, and the author, calpionellids, i. e. microplancton). Stable isotope analyses: $\partial^{13}C$ (whole rock) and $\partial^{18}O$ were provided by H. Hubberten, Potsdam. The different analyses were made on the same samples, in order to ascertain a direct correlation of the results.

The Jurassic/Cretaceous boundary was provisionally placed at the base of the Jacobi ammonite Zone following the "Colloque sur la limite Jurassique-Crétacé, Lyon/Neuchâtel 1973" [5]; this level corresponds very closely to the base of calpionellid Zone B [11, 12]. Due to a strong paleobiogeographic differentiation, this boundary is, however, extremely difficult to correlate; the results of certain authors differ by nearly 3 ammonite zones. The controversy between Jeletzky and Zeiss [7], concerning the supposed diachronism of the ammonite genus *Durangites* across the Atlantic illustrates this uncertainty. Due to the great proportion of endemic genera in Mexican ammonite faunas, the question remained open. But Mexican calpionellids correlate well with the Mediterranean Basin; the absence of endemic species [4, 14] was confirmed by our observations. We were thus able to confirm the correlation of Zeiss [7] who considered *Durangites* as an isochronous marker, ranging up into the basal part of Zone B also in Mexico.

Tithonian calpionellids are, however, extremely rare and badly preserved in Mexico. Therefore the presence of associations indicating calpionellid Zone A remains uncertain. A hiatus of the Tithonian can nevertheless be excluded and the near-to absence of *Crassicollaria* and of *Saccocoma* can be attributed to paleobiogeographic factors [3]. Datable faunas appear only within Zone B, in the lowermost Berriasian. This invasion of E Mexico by calpionellids corresponds to a characteristic change from more detrital, sometimes phosphatic micrites rich in radiolarians to typical calpionellid limestones. In the same time various Mediterranean ammonite genera appear (Fig. 1). The biogeographic turnover is coupled with a rise of $\partial^{13}C$ values. A coeval tendency to higher $\partial^{18}O$ values seems to indicate a slight cooling. As to clay minerals, the existence of a chlorite spike covering the lower part of the Berriasian up into calpionellid Subzone D 2 is another indicator for important oceanographic changes occurring in the Lower Berriasian, probably related to a eustatic sea-level rise.

Unfortunately, the improved correlation between Mexico and the Mediterranean Basin does not resolve the problem of the Jurassic/Cretaceous boundary, as the first correlatable level lies slightly above that boundary. But the study of Mexican calpionellid faunas has brought another interersting result, important for sequence stratigraphy. Several successive faunal spreads occurred in the Berriasian, each of them extending farther West. At San Pedro del Gallo, nearly 300 km W of the Sierra Madre Oriental, the oldest calpionellid faunas belong to the mid-Berriasian Zone C, and in the state of Oaxaca, calpionellids do not appear before the the late Berriasian Zone D. There seem to be several closely spaced faunal spreads within this zone, at least partly related to eustatic highstands, as shown by calpionellid-bearing intercalations in carbonate platform successions in the Jura Mountains and the N Caucasus (Baksan section).

The example presented above shows that the use of different fossil groups may diminish, but not overcome the problem of paleobiogeographic barreers. Therefore, another example, where stable isotopes play a more important rôle shall be briefly discussed here:

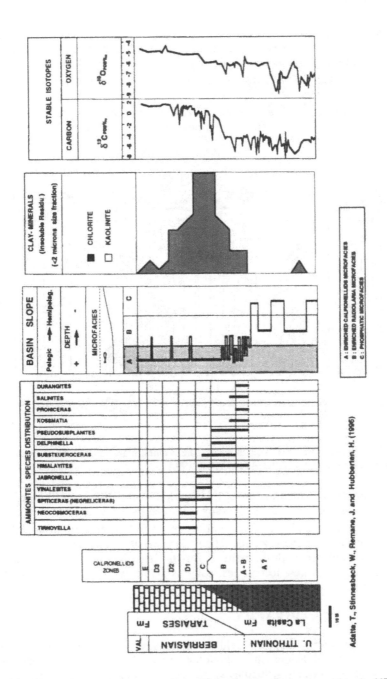

Figure 1 Multidisciplinary stratigraphy of the Tithonian-Berriasian transition in NE Mexico: stratigraphic distribution of key ammonite genera, of microfacies, clay-mineral associations, and of ∂^{13} C and ∂^{18} O values. After Adatte *et al.* 1996.

THE STABLE ISOTOPE SHIFT AT THE VALANGINIAN/HAUTERIVIAN BOUNDARY

An important positive $\partial^{13}C$ excursion covering more or less the Valanginian to lower Hauterivian C. oblongata nannoplankton Zone, was first discovered in 5 localities in the Lombardian Basin of the Southern Alps and could be calibrated with magnetostratigraphy in one of them [8]. The existence of that $\partial^{13}C$ excursion in several DSDP sites in the Western North Atlantic, the Gulf of Mexico and the Central Pacific [8] proves that we deal here with a truly global event. Recently this $\partial^{13}C$ excursion was also discovered in two localities of NE Mexico [1] so that it is now also accessible to the field geologist on the other side of the Atlantic.

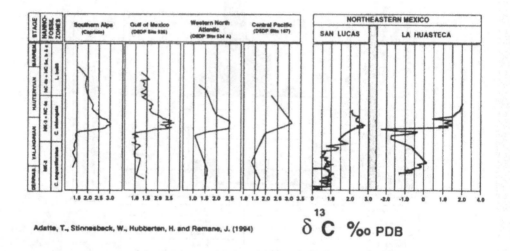

Adatte, T., Stinnesbeck, W., Hubberten, H. and Remane, J. (1994) $\delta\ ^{13}C\ \%_0\ PDB$

Figure 2 Evolution of $\partial^{13}C$ values across the Valanginian/Hauterivian boundary in different profiles all over the world. After Adatte *et al*. 1993.

Unlike biostratigraphic correlations, this isotopic correlation is not limited by facial or paleobiogeographic barreers. Its accuracy is limited by the duration of the "event", but the main problem is that similar $\partial^{13}C$ shifts are known from the Aptian/Albian and the Cenomanian/Turonian transition [8]. The isotope event is repetitive, successive events can only be distinguished with the help of fossils. It is thus the combination of classical biostratigraphy with new techniques, such as stable isotope analyses which allows to make progress.

MAGNETOSTRATIGRAPHY

Reversals of the Earth's magnetic field are in many respects ideal marker events. They are truly global, and with a duration of about 5000 years [9], they are instantaneous compared to other geologic processes, such as stable isotope excursions. With the help of oceanic magnetic anomalies, a detailed Geomagnetic Polarity Time Scale (GPTS) has been established for the period extending from the Callovian/Oxfordian up to the Recent. If magnetic polarity chrons of the GPTS can be correctly identified in the sections under study, they will thus provide a means of very precise global correlation.

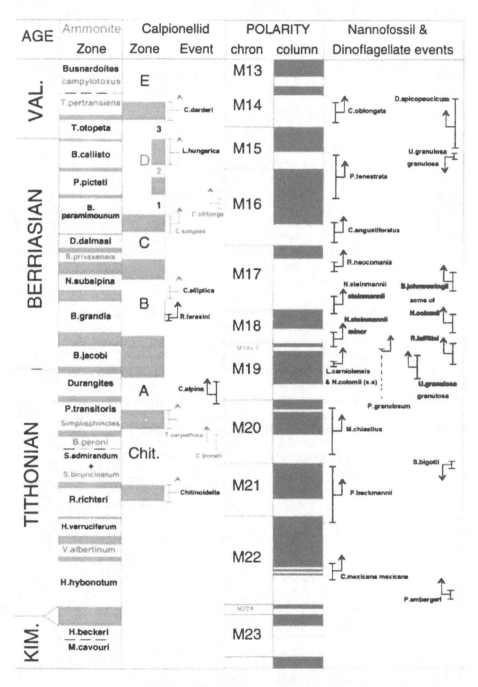

Figure 3 Range of chronostratigraphic variation of biostratigraphic calpionellid and nannoplancton events, determined with the help of magnetostratigraphy. After Ogg *et al.* 1991, position of the Tithonian-Berriasian boundary modified

The problem is again the repetitivity of events. The direct comparison of magnetic zones observed in the field with the GPTS is only possible in continuous sedimentary successions with a more or less constant sedimentation rate. Even then a distortion of the "fingerprint" is inevitable. Therefore it is difficult to match magnetic zones with the magnetic polarity chrons, especially if we consider that different parts of the GPTS may be quite similar to one another, such as the Upper Oxfordian to Kimmeridgian interval to the Valanginian/Lower Hauterivian [cf. 9] or the Oligocene/Miocene transition to the Middle Miocene [cf. 6].

The situation changes dramatically, as soon as the successions can be dated by fossils. Even an approximate datation will normally allow the correct identification of all or most of the magnetic polarity chrons in a section [9, 10]. The most surprising result is that after having been determined with the help of biostratigraphy, magnetic zones can serve in return to measure the degree of diachronism of biostratigraphic boundaries (Fig. 3), using the boundaries of magnetic zones as isochronous reference [9]. One could even go one step farther and use magnetostratigraphy to calibrate phyletic first appearances and extinction events: Obviously the extremes of the variation (and not the means, as one might be inclined to believe) provide the best approximations: the first of all biostratigraphic occurences will be closest to the real phyletic appearance, and the last occurrence closest to the moment of exctinction. In other words, magnetostratigraphy helps to progress from biostratigraphy to biochronology.

The unique possibilities of Paleogene and Neogene chronostratigraphy due to the detailed calibration of magnetochrons and radiometric ages with zones based on planktic foraminifera and on nannoplankton can best be illustrated by the GSSP for the base of the Neogene, which was voted by ICS last year and ratified by IUGS on this Congress. Two figures taken from the submission presented for vote to ICS by the Paleogene/Neogene Boundary Working Group [13], are reproduced here (Figs 4, 5) The overlook (Fig. 4) shows the magneto, litho, and biostratigraphy of the type-section. Magnetic polarities could not be determined throughout the section, but the observed boundaries could be attributed to polarity chrons of the GPTS. Note that the base of the Neogene was defined to coincide in the type-section with the base of the magnetic zone thought to correspond with the base of polarity chron C6Cn2n.

It should, however, be stressed that the existence of a reliable GPTS based on oceanic anomalies is the decisive prerequisite for this kind of approach. Such a standard does not exist for the time before the Callovian/Oxfordian, which is the age of the oldest oceanic crust. The possibilities to use magnetostratigraphy in older rocks are thus much more limited. If a given event occurs in sediments of different polarity in different sections, all what this proves is that the respective levels cannot be of the same age. If a biostratigraphic zone corresoponds to three magnetic zones in one section and only to two in another one, the latter is without any doubt less complete. But, unlike the examples presented above, it is impossible to correlate magnetic zones from one section to another. Both sections may well be incomplete at different levels.

CONCLUSIONS

So, what can be learned from these examples of multidisciplinary chronostratigraphy? First of all that the integration of new methods of correlation, based on magnetic reversals, stable isotopes and other geochemical signals reflecting global events, as well as the detection of Milankovich cycles in the sedimentary record are of vital importance. We have seen that these methods can provide truly interregional correlations, and that some of

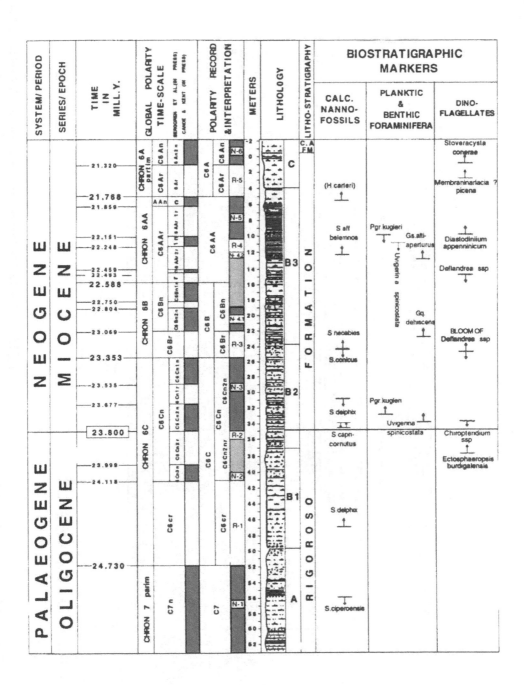

Figure 4 Biostratigraphy, radiometric ages, magnetostratigraphy, and lithostratigraphy, of the Lemme-Carrosio section (Italy), type-section for the Paleogene/Neogene boundary. After Steininger *et al.* 1994.

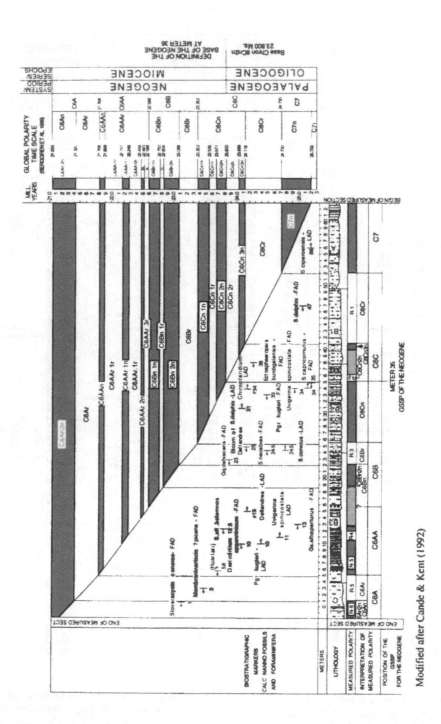

Figure 5 Plot of the GPTS versus the paleomagnetic polarity sequence, lithology and major biostratigraphic events of the Lemme-Carrosio section. After Steininger *et al.* 1994.

them - especially magnetostratigraphy, are also very accurate. But it has to be recalled that the new methods cannot simply replace the old ones - i. e the dating of rocks by fossils or radioisotopes, because only these rely on irreversible processes - radioactive decay or organic evolution, and only these allow to date rocks in terms of numerical or relative ages. The new methods of correlation use, in the contrary, repetitive events. Therefore they have to be combined with methods of dating in order to determine reliable tie points for correlations. Then, however, they will increase the range and the accuracy of classical biostratigraphic methods of dating.

So what is multidisciplinary chronostratigraphy? It is the combination of all available methods in the study of one problem, and this means cooperation and team work.

REFERENCES

1. T. Adatte, W. Stinnesbeck, H, Hubberten and J. Remane. Isotopos estables en el Valanginiano del Noreste de México, su relación con cambios a nivel global, *3º Congr. Nacion. Geoquím.* **1993**: 79-85 (1993).
2. T. Adatte, W. Stinnesbeck and J. Remane. The Jurassic-Cretaceous boundary in Northeastern Mexico, confrontation and correlations by microfacies, clay minerals mineralogy, calpionellids and ammonites,*Geobios*, Mém. spécial **17**: 37-56 (1994).
3. T.Adatte, W. Stinnesbeck, J. Remane and H. Hubberten. Paleogeographic setting of Center-East Mexico at the Jurassic/Cretaceous boundary, correlation with NE-Mexico, *Mitt. Geol.- Paläont. Inst. Univ. Hamburg* **77**: 379-393 (1996).
4. F. Bonet. Zonificación microfaunística de las calizas cretácicas del Este de México, *Bol. Asoc. Mex. Geol. Petrol* **8**: 389-488 (1956).
5. Colloque sur la limite Jurassique-CrétCé, Lyon, Neuchâtel Septembre 1973: Discussions sur la position de la limite Jurassique-Crétacé, *Mém. Bureau Rech. Géol. et Min.* **86**: 379-393 (1975).
6. E. A. Hailwood. Magnetostratigraphy, *Geol. Soc.*, Spec. Report **19**, 84 p., Blackwell (1989)
7. J. A. Jeletzky. Jurassic-Cretaceous boundary beds of Western and Arctic Canada and the problem of the Tithonian-Berriasian stages in the Boreal Realm., *in* : G. E. G. Westermann (ed.): Jurassic-Cretaceous biochronology & biogeography of North America, *Geol. Assoc. Canada*, spec. paper **27**: 175-276, including Comments by A. Zeiss: 250-253 (1984).
8. A. Lini, H. Weissert and E. Erba. The Valanginian carbon isotope event: a first episode of greenhouse climate conditions during the Cretaceous, *Terra nova*, 4/3:374-384 (1992).
9. J. G. Ogg, R. W. Hasenyager, W. A. Wimbledon, J. T. E. Channell and T. J. Bralower. Magnetosstratigraphy of the Jurassic-Cretaceous boundary interval - Tethyan and English faunal realms,*Cretaceous Research* **12**: 455-482 (1991).
10. J. G. Ogg and W. Lowrie. Magnetostratigraphy of the Jurassic-Cretaceous boundary, *Geology* **14**: 547-550 (1986).
11. J. Remane. Les calpionelles dans les couches de passage Jurassique-Crétacé de la fosse vocontienne, *Trav. Lab. Géol. Fac. Sci. Grenoble* **39**: 25-82 (1963).
12. J. Remane. Calpionellids. - *in* : H. Bolli, J. B. Saunders and K. Perch-Nielsen (eds.) Plankton Stratigraphy, *Cambridge Univ. Press* : 555-572 (1985).
13. F. F. Steininger, M. P. Aubry, M. Biolzi, A. M. Borsetti, F. Cati, R. Corfield, R. Gelati, S. Iaccarino, C. Napoleone, F. Rögl, R. Roetzel, S. Spezzaferri, F. Tateo, G. Villa and D. Zevenboom. Proposal for the Global Stratotype Section and Point (GSSP) for the base of the Neogene (the Paleogene/Neogene boundary), *Inst. for Paleont., Univ. of Vienna*, 41p.
14. M. Trejo. Tintínidos mesozoicos de México (taxonomía y datos paleobiológicos), *Bol. Asoc. Mex. Geol. Petrol.* **27**: 329-449 (1976).

Proc. 30ᵗʰ Int'l. Geol. Congr., Vol. 11 , pp. 11-19
Wang Naiwen and J. Remane (Eds)
© VSP 1997

Some Constraints on the Phanerozoic Timescale

F. GRADSTEIN, Saga Petroleum, N-1301 Sandvika, Norway; J. OGG, Department of Earth
and Atmospheric Sciences, Purdue University, West Lafayette, Indiana 47907, U.S.A.; F.
AGTERBERG, Geological Survey of Canada, Ottawa, Ontario, K1A OE8, Canada; J.
HARDENBOL, Global Sequence Chronostratigraphy Inc., 826 Plainwood, Houston, TX
77079, USA; and P. van Veen, Norsk Hydro Research - Production, N-5020 Bergen, Norway.

Abstract

Stratigraphic and geomathematical interpolation methods are needed to achieve the Phanerozoic time scale.
Currently, only the Mesozoic part has estimates of uncertainty attached to the age of the 31 successive stage
boundaries. Constraints on the current Phanerozoic scale are both of a stratigraphic and geochronologic nature,
as outlined for several stage boundaries in the Paleozoic and Mesozoic. Agreement, not only on the
stratigraphic definition of stages, but also on employment of time scale methods will enhance the consensus
character of the geologic time scale and stability in earth science applications.

Keywords: Geological time scale, Geochronology, Radiometrics, Phanerozoic, Cambrium, Triassic, Seafloor
spreading.

PHANEROZOIC TIME SCALE

The ideal geologic time scale makes use of precise and standardized age estimates at successive stage
boundaries for the Phanerozoic, although with the current state of knowledge, we are still far from that
goal. During the last decade, significant progress has been made in defining Phanerozoic stage
boundaries in a stratigraphic sense (J. Remane, pers. comm., 1996). Nevertheless, many stage limits,
particularly in the middle part of Phanerozoic leave to be desired, or are being re-defined to ensure
better biostratigraphic and geochronologic calibration. Apart from fundamental chronostratigraphic
issues, there is the problem of sufficient and stratigraphically meaningful age dates, and their
intercalibration to a common standard.

For most of the Phanerozoic scale, interpolation methods are used where precise and accurate age dates
on stage boundaries are lacking. The Mesozoic and Cenozoic geologic time scale builds on the observed
and interpolated ties between (1) radiometric dates, biozones and stage boundaries, and (2) between
biozones and magnetic reversals on the seafloor and in sediments. The Paleozoic scale is a loose
assembly of sparse isotopic age estimates for selected biostratigraphic levels, that are far apart.

Figure 1 provides a summary of current time scale methods. It is beyond the scope of this brief review
to reiterate methodological details, many of which have been addressed in studies cited below. The
outline in figure 1 is meant as a simple and quick overview of common methods employed in furnishing
a geologic time scale. It is our philosophy that a better understanding of methods and constraints in
establishing the scale will enhance consensus, to the benefit of consistent applications in earth sciences.

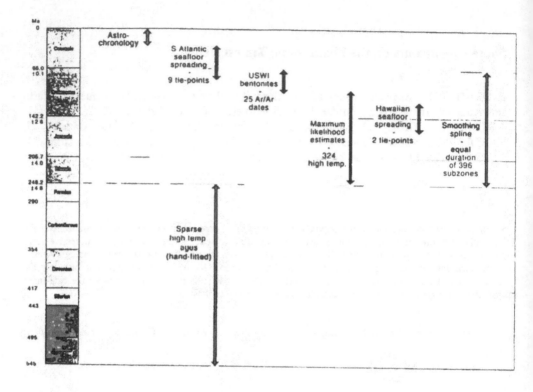

Figure 1. Methods to achieve the geologic time scale in figure 2.

The Cenozoic time scale is primarily calibrated from biostratigraphic correlations to magnetic polarity chrons, which in turn are scaled according to marine magnetic anomaly profiles from the South Atlantic seafloor, pinned to a selected set of 9 Ar/Ar radiometric ages [2]. Interpolation, using constancy of seafloor spreading between radiometrically constrained profile segments, assigns ages to magnetic polarity chrons which, in turn calibrates zonal events and stage boundaries. The combination of magnetochronology and astrochronology presently refines the late Neogene time scale in a continuous and linear manner for the last 6 m.y. [20]. Efforts are underway to extend the continuous astrochronologic scale back into Oligocene (N. Shackleton, pers. comm., 1996)

The unifying interpolation concept of seafloor spreading in parts of the Cenozoic scale is not feasible in the Mesozoic, because magnetic anomaly profiles on the seafloor only extend back to the middle Jurassic and the middle Cretaceous lacks magnetic reversals. Therefore, whereas portions of the Mesozoic time scale can now be exactly determined by a combination of precise radiometric ages, many published in the last five years on biostratigraphically constrained sections [e.g. 25]) for the Upper Cretaceous of the United States Western Interior], the majority of the stage boundaries have been assigned ages through geological and mathematical interpolation methods. Two such methods, both highly sensitive to the varying precision of stratigraphic data, include maximum likelihood and spline interpolations, which together preserve an estimate of uncertainty for the 31 Mesozoic stage

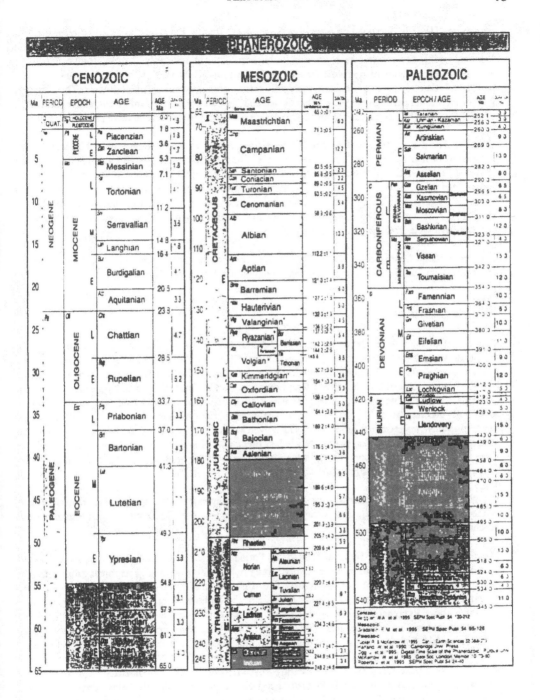

Figure 2. Phanerozoic time scale (slightly modified after Gradstein and Ogg, 1996).

boundaries [1]. The interpolation method presently used for the Paleogene and early Neogene stages does not preserve uncertainties. Since the chronostratigraphic and geochronologic compilation for the Paleozoic [18], several studies have been published that use relatively precise isotope datings to improve Paleozoic chronology below the Carboniferous [33, 34]. Nevertheless, there remains a general paucity of relatively precise and stratigraphically meaningful age dates in the Paleozoic. This, together with problems in inter-regional stage assignments, and the absence of standard zonations for some parts of the Paleozoic (R. Cocks, pers.comm.,1996) makes the Paleozoic time scale more uncertain.

We have incorporated the new Cenozoic, Mesozoic and Paleozoic age assignments in the geologic time scale chart of figure 2, with reference to its sources. The Cenozoic scale is after Berggren et al. [2] and Hilgen [20], and the Mesozoic one after Gradstein et al. (11, 12]. The Paleozoic time scale is constructed from Harland et al. [18], with the pre-Carboniferous updated by more precise isotopic ages for stage boundaries [33, 34]. A recent magnetochronology for the Phanerozoic has been published by Ogg [28]. A virtually identical and coloured version of the Phanerozoic time scale, in A3 format, is currently available from the journal 'Episodes'.

The consistent Cenozoic-Mesozoic time scale forms the basis for an extensive suite of chronostratigraphic and sequence-stratigraphy compilations for Europe [17].

SOME CONSTRAINTS

The advent of radiometric dating techniques with less than 1% analytical error, as demonstrated in recent Cretaceous and Triassic U/Pb and Ar/Ar dates [24, 25] furnishes a major challenge to both biostratigraphy and interpolation mathematics. Even the most detailed biostratigraphic scheme probably has no biozonal units of less than 0.5- 1.0 myr duration, not to speak of the actual precision in dating a particular 'stratigraphic piercing' point, for which an U/Pb age estimate would be available with an analytical uncertainty of 0.1 to 0.5 myr. The combination of such dates with high-resolution biostratigraphy, magnetostratigraphy or Milankowitch cyclicity is a major challenge to time scale studies. Similarly, combination of analytically less precise K/Ar and much more precise Ar/Ar or U/Pb dates in statistical interpolations, creates a strong bias towards the latter, despite the fact that both may have equal litho- and chronostratigraphic precision.

Below, we will present examples of the resulting complexities in stratigraphic reasoning and mathematical interpolations, using recent developments in the Phanerozoic stratigraphy and geochronology.

The 'Working Group on the Terminal Proterozoic Period' chaired by A. Knoll (Harvard, Mass., USA) is erecting a relatively detailed bio- and isotope stratigraphy framework and geochronology for this 50+ m.y. period below the Cambrian, using new data from Australia and Namibia. A GSSP (Global Boundary Stratotype Section and Point) for the base of the Terminal Proterozoic is considered in a cap carbonate above Varanger tillites in southern Australia, that from correlations to tuffs with U/Pb dates on zircons in Namibia may be near 570 Ma old. [14]. Arguments are now emerging for a slight upwards revision of the age of base Cambrian, the Period immediately overlying the Terminal Proterozoic [14, 7, 6]. The GGSP for the base of the Cambrian in Newfoundland, marked by the appearance of a trace fossil assemblage bearing Phycodes-pedum, can be correlated with strata in N. Siberia, assigned into a Nemakit-Daldynian age. Volcanic breccia, slightly below Phycodes-pedum have yielded a U/Pb age of 543.6 ± 0.24. In Namibia, this trace fossil assemblage appears near the base of the Nomtsas Formation, above the Spitskopf-Nomtsas erosional unconformity that includes the Precambrian - Cambrian boundary. The boundary is bracketed by U/Pb ages in ashes of 539.4 ± 1 Ma and 543.3 ± 1 Ma. Hence the base of Cambrian may be at, or slightly above, 543 Ma.

The Subcommission of Permian Stratigraphy under the International Stratigraphic Commission (see also Jin Yugan in number 28, June 1996 of the subcommissions newsletter "Permophiles") is refining a proposal for Permian stage and series classification, incorporating new insights from China, Kazakhstan, Russia and North America. The 'early' Permian Cisuralian subperiod incorporates the stages Asselian, Sakmarian, Artinskian and Kungurian. The 'middle' Permian Guadalupian subperiod consists of the stages Roadian, Wordian and Capitanian, whereas the youngest Permian subperiod, the Lopingian incorporates the two successive stages Wuchiapingian and Changhsingian. In the same newsletter cited, A. Klets et al. quote new U/Pb (SHRIMP) dates from tuffs in the lower Permian of Russia and eastern Australia. Although it is not clear how the dates fit in the above stage classification, Sakmarian through Kungurian stages might shift down 5-10 m.y., pending stratigraphic and mathematical analysis of calibrations, dates and errorbars. In the confusion around uppermost Permian stratigraphy, global correlation of the Illawarra geomagnetic reversal from the Tatarian of Russia to China, Australia and other Perm regions is important [M. Menning pers. comm., 1996; 23].

In the ongoing debate on the age of the Paleozoic/Mesozoic boundary and radiometric calibrations, P.R. Renne and colleagues [29, 30] are radiometrically correlating the (?) oldest flows in the Siberian flood basalts and minor tuffs virtually at the Permian/Triassic boundary in candidate Global Stratotype sections and Point (GSSP) for this boundary in southern China. When adjusting the MMhb-1 standard for Ar/Ar to the Fish Canyon standard for U/Pb, the P/T boundary is at 250 Ma, which has the benefit of being a convenient number to use and remember. Although no details are given for the (undoubtedly relatively small) stratigraphic uncertainty of the age estimate, analytical uncertainty appears to be on the order of less than 0.95 m.y. The challenge will be to stratigraphically calibrate the Siberian flood basalts to the GSSP stratigraphy in China, and to estimate on a linear scale how close best age estimates are to the GSSP. Independent calibration of the Ar/Ar standard to the same orbital cycle scale as achieved for U/Pb [31] is also desirable.

An important anchor point in the Triassic is dating of the Ladinian/Anisian boundary in the Grenzbitumen horizon in Switzerland [19; items 309 and 310 in 31]. Both K/Ar and Ar/Ar dates on alkali feldspars were done, from tuff layers in the basal part of the lowermost Ladinian *Nevadites* (tethyan) ammonite zone [3]. The best dates are derived from homogenous and clear, high-sanidine feldspars (type G), which average at 232 ± 9 Ma (2s) for K/Ar and 233 ± 7 Ma (2s) for Ar/Ar, from which Hellmann and Lippolt [26]) estimate the boundary to be 232 ± 9 Ma (2s). Odin [26] suggests 232.7 ± 4.5 Ma (2s) for K/Ar and 232 Ma plateau age for Ar/Ar in the waterclear, high sanidine feldspars (NDS196). This best estimate was used in interpolating the Early/Middle Triassic boundary at 234.3 ± 4.6 Ma (2s), [11, 12].

In May 1995, Brack et al. [4] reported single-grain zircon U/Pb age dates from tuffaceous layers associated with the Anisian/Ladinian boundary interval in sections near Bagolino in northern Italy [see also 24]. This region is a proposed candidate for the GSSP of the base of the Ladinian stage, although disagreements on the precise level have not yet been resolved. A suite of seven zircons from a thin crystal tuff in the lower part of the *Secedensis [Nevadites]* ammonite zone yielded a weighted Pb/U mean age of 241.0 ± 0.5 Ma. This same bed can be traced to equivalent tuffs in the Grenzbitumenzone at Monte San Giorgio in southern Switzerland [3]. As summarized above, sanidine feldspars from these tuffs in the Grenzbitumenzone had previously yielded a mean K/Ar age of 233 ± 4.5 (2s) and a plateau age of approximately 232 ± 4.5 Ma (2s). However, Brack et al. [4] report that preliminary zircon ages from this same Grenzbitumenzone are consistent with their age from Bagolino, and also report that the regional burial and maturation history favours the zircon ages to be stratigraphically reliable. An upper Etalian tuff in New Zealand, biostratigraphically correlated to Upper Anisian recently yielded Ar/Ar ages near 242.8 Ma [32], compatible with the new Grenzbitumen age.

Incorporation of the new radiometric date for this boundary interval and other zircon ages [4] for overlying middle Ladinian tuffs (238.8 ± 0.4 Ma in the middle of *Gredleri* Zone; 237.7 ± 0.5 Ma in the middle of *Archelaus* Zone) into the maximum-likelihood database [12], and new spline computations imply a longer time span for the Ladinian Stage (8.3 instead of 6.9 m.y.), and a greatly reduced extent of the Anisian Stage (3.1 instead of 7.4 m.y.). However, Brack et al. [4] recommend placement of the *Nevadites* Zone into the Anisian Stage, rather than its traditional placement in the Ladinian, which would result in a stratigraphically slightly younger Ladinian/Anisian boundary.

It is curious to note that the potential changes in duration of the two successive stages greatly reduce the duration of Anisian ammonite zones over Ladinian ones. An unresolved issue of the new geochronologic information is that the 600 Latemar carbonate platform cycles in the Middle Triassic of the southern Alps, Italy, appear to be formed during a shorter period of time than previously assumed. The new stage durations cast doubt on recognition of orbital cycle frequencies in these carbonates [5].

In a similar vein, Milankovitch cyclicity may not be applicable to the 'van Houten' cycles in the uppermost Triassic of the Newark Basin [35]. These cycles were previously used to approximate the timespan between the disappearance of vesicate pollen, the palynologically assigned Triassic/Jurassic boundary and the onset of basalt deposition in the Newark Basin [9]. In the present Triassic time scale, the lacustrine cycles may not represent single Milankovitch cycles.

The Oxfordian through earliest Aptian magnetic anomaly record of the east Pacific Ocean, coupled with magneto-biostratigraphic correlations provides a powerful means for relative scaling of the associated stages [22, 21, 16, 11, 12]. In the absence of more reasonable constraints, constancy of seafloor spreading was generally utilized to interpolate Late Jurassic through Early Cretaceous ages. The ages for stage boundaries derived from a constant spreading rate assumption must be reconciled with the radiometric database and associated maximum likelihood age estimates. The Mesozoic time scale of figure 2 combines the seafloor spreading scaling assumption with independent maximum likelihood estimation for the ages of the Oxfordian through Barremian stage boundaries. In the final time scale, limited weight was given to interpolated ages from constant seafloor spreading. Figure 3 illustrates Late Jurassic through Early Cretaceous Pacific spreading rates implied by different time scales. We consider it significant that the Berriasian spreading rate is significantly slower than the Valanginian one, suggestive of first-order correlation to observed global onlap in the Valanginian, following major regression [16, 17]. More detailed studies may resolve a similar first order 'causal coincidence' for trends in Jurassic spreading rates and oflap/onlap cycles, and provide geological sense to geochronology.

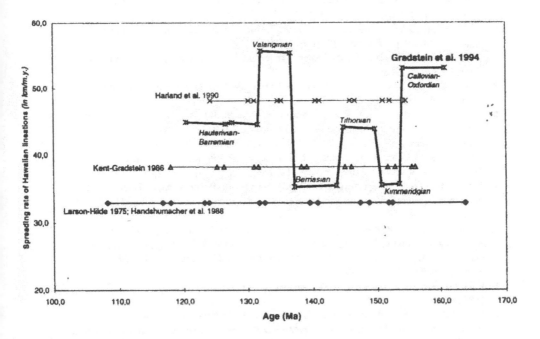

Figure 3. Late Jurassic through Early Cretaceous Pacific spreading rates, implied by different time scales.

ACKNOWLEDGEMENT

The senior author is pleased to acknowledge the support from the exploration research section of Saga Petroleum in completing this study and drafting illustrations.

REFERENCES

1. Agterberg. Estimation of the Mesozoic Geological Time Scale. *Math. Geology* 26 (7), 857- 876 (1994).
2. W.A. Berggren, D.V. Kent, C.C. Swisher III and M.P. Aubry. A revised Cenozoic geochronology and chronostratigraphy. *In* Berggren et al., eds., *Geochronology, Time Scales' and Global Stratigraphic Correlation;* Tulsa. *SEPM Special Publication* 54, 129 -212 (1995).
3. P. Brack, P. and H. Rieber, 1994. The Anisian/Ladinian boundary: Retrospective and new constraints. *Albertiana* 13, p. 15 - 36.
4. P. Brack, H. Rieber, and R. Mundi. The Anisian/Ladinian boundary interval at Bagolino (southern Alps, Italy): I. Summary and new results on ammonoid horizons and radiometric age dating. *Albertiana* 15: 45-56 (1995).

5. P. Brack, R. Mundil, F. Oberli, M. Meier, and H. Rieber. Biostratigraphic and radiometric age data question the Milankovitch characteristics of the Latemar cycles (Southern Alps, Italy). *Geology* 24, p. 371-375 (1996).

6. M.D. Brasier, J.W. Cowie, and M. Taylor. Decision on the Precambrian-Cambrian boundary stratotype. *Episodes* 17, p. 3-8 (1994).

7. M.D. Brasier. U/Pb dates for the base of the terminal Proterozoic and Cambrian. *Episodes* 19, no's 3 & 4 (1996).

8. J.W. Cowie and M.G. Bassett. 1989 Global Stratigraphic Chart. Ottawa, International Union of Geological Sciences (1989).

9. S.J. Fowell and P.E. Olsen. Time calibration of Triassic/Jurassic microfloral turnover, eastern North America. *Tectonophysics* 222, 361-369 (1993).

10. F.M. Gradstein, F.P. Agterberg, M.P. Aubry, W.A. Berggren, J.J. Flynn, R. Hewitt, D. Kent, K. Klitgord, K.G. Miller, J.D. Obradovich, J.G. Ogg, D.R. Prothero, and G.E.G. Westermann. Sea level history. *Science* 214, 599 - 601 (1988).

11. F.M. Gradstein, F.P. Agterberg, J.G. Ogg, J. Hardenbol, P. van Veen, J. Thierry, and Z. Huang. A Mesozoic time scale. *Journal of Geophysical Research* 99, 24051 - 24074 (1994).

12. F.M. Gradstein, F.P. Agterberg, J.G. Ogg, J. Hardenbol, P. van Veen, J. Thierry, and Z. Huang. A Triassic, Jurassic, and Cretaceous time scale. *In* W. Berggren et al., eds., *Geochronology, Time Scales and Global Stratigraphic Correlation*; Tulsa. *SEPM Special Publication* 54, 95 - 126 (1995).

13. F.M. Gradstein and J.G. Ogg. A Phanerozoic time scale. *Episodes* 19, (1 & 2), 3-5 plus chart (1996).

14. J.P. Grotzinger, S.A. Bowring, B.Z. Saylor, and A.J. Kaufman. Calibrating the terminal Proterozoic time scale. *30th Intern. Geol. Congress*, Beijing, *Abstract volume* 2, 47 (1996).

15. D.W. Handschumacher, W.W. Sager, T.W.C. Hilde, and D.R. Bracey. Pre-Cretaceous evolution of the Pacific plate and extension of the geomagnetic polarity reversal time scale with implications for the origin of the Jurassic "Quiet Zone". *Tectonophysics* 155, 365-380, (1988).

16. B.U. Haq, J. Hardenbol, and P.R. Vail. Mesozoic and Cenozoic chronostratigraphy and cycles of sea-level change, in *Sea-level Changes-An integrated approach, SEPM Special Publ.* 42, 71-108 (1988).

17. J. Hardenbol, J. Thierry, J. Rey, P.R. Vail, T. Jacquin, P.C. De Graciansky, and M.B. Farley, *Mesozoic and Cenozoic sequence chronostratigraphy of European Basins*, in prep.

18. W.B. Harland, R.L. Armstrong, A.V. Cox, L.E. Craig, A.G. Smith, and D.G. Smith. A Geologic Time Scale 1989. *Cambridge University Press* (1990).

19. K.N. Hellmann and H.J. Lippolt. Calibration of the Middle Triassic time scale by conventional K-Ar and Ar-Ar dating of alkali feldspars, *J. Geophys.*, 50, 73-86 (1981).

20. F.J. Hilgen. Extension of the astronomically calibrated (polarity) time scale to the Miocene/Pliocene boundary. *Earth and Planetary Science Letters* 107, 349-368 (1991).

21. D.V. Kent and F.M. Gradstein. A Jurassic to Recent Chronology. In, *The Geology of North America, Volume M, The Western North Atlantic Region*, edited by P.R.Vogt and B.E. Tucholke, *Geological Society of America*, Boulder, CO (1986).

22. R.L. Larson and T.W.C. Hilde A revised time scale of magnetic reversals for the Early Cretaceous and Late Jurassic, *J. Geophys. Res.* 80, 2586-2594 (1975).

23. M. Menning. A numerical time scale for the Permian and Triassic Periods: An integrated analysis. In, Scholle, P. Pery, T., and Ulmer-Scholle, D., (eds): *The Permian of northern Pangea* 1, *Springer Verlag*, Berlin, 77-97 (1995).

24. R. Mundil, P. Brack, M. Meier, H. Rieber and F. Oberli. High-resolution U-Pb dating of middle Triassic volcaniclastics: Time-scale calibration and verification of tuning parameters for carbonate sedimentation. *Earth and Planetary Science Letters* 141, 137-151 (1996).

25. J.D. Obradovich. A Cretaceous time scale. In, *Evolution of the Western Interior Basin, Geological Association of Canada Special Paper* 39, ed. W.G.E. Caldwell, 379-396 (1993).

26. G.S. Odin (ed.). Numerical Dating in Stratigraphy 1, 2. *J. Wiley & Sons*, Chichester, U.K (1982).

27. G.S. Odin and C. Odin. Echelle Numerique des Temps Geologiques. *Geochronologie*, 35, 12 - 20 (1990).

28. J.G. Ogg. Magnetic polarity time scale of the Phanerozoic. *In* Ahrens, T., ed., *Global Earth Physics: A Handbook of Physical Constants*; Washington, *American Geophysical Union Reference Shelf* 1, 240-270 (1995).

29. P.R. Renne and A.R. Basu. Rapid eruption of the Siberian Traps flood basalts at the Permo-Triassic boundary, *Science* 253, 176-179 (1991).
    30. P.R. Renne, Z. Zichao, M. Richards, M. Black, and A. Basu. Synchroneity and causal relations between Permian-Triassic boundary crises and Siberian flood volcanism. *Science* 269, 1413 - 1416 (1995).

31. P.R. Renne, A.L. Deino, R.C. Walther, B.D. Thurrin, C.C. Swisher, T.A. Becker, G.H. Curtis, W.D. Sharp, and A.R. Jaouni. Intercalibration of astronomical and radioisotopic time. *Geology* 22, 783 (1994).

32. G.J. Retallack, P.R. Renne, and D.L. Kimbrough. New radiometric ages for Triassic floras of Southeast Gondwana. In: Lucas, S.G. and Morales, M., eds. *The Nonmarine Triassic. New Mexico Museum of Natural History & Science Bulletin* 3, 415 - 418 (1993).

33. J. Roberts, J. Claoue-Long, and P.J. Jones. Australian Early Carboniferous Time. *In* Berggren et al., eds., *Geochronology, Time Scales and Global Stratigraphic Correlation*; Tulsa, *SEPM Special Publication* 54, 24 - 40 (1995).

34. R.D. Tucker and W.S. McKerrow. Early Paleozoic chronology: A review in light of new U- Pb zircon ages from Newfoundland and Britain. *Canadian Journal of Earth Sci.* 32, 368 - 379 (1995).

35. P. van Veen. Time calibration of Triassic/Jurassic microfloral turnover, eastern North America - Comment. *Tectonophysics* 245, 93-95 (1995).

Proc 30ᵗʰ Int'l. Geol. Congr., Vol. 11, pp. 21-27
Wang Naiwen and J. Remane (Eds)
© VSP 1997

TBO-STRATOTYPE SYSTEM AND NON-SMITH STRATIGRAPHY

Some considerations on a strategy how stratigrahpy can jump again to the forefront of geoscience

WANG NAIWEN

Institute of Geology and Lithosphere Research Center. Chinese Academy of Geological Sciences. Beijing 100037. China

Abstract

The principal mission of stratigraphy is to establish the temporal succession of all environments of the Earth in the geologic past. The stratotypes, either recently erected or yet planned to be defined, are only suitable for one environment: epicontinental/marginal seas. They can hardly apply to other environments like land areas and oceans which occupy/occupied the greatest part of the Earth's surface and can produce more or less continuous sedimentary and fossil records as well. The author, therefore, suggests that stratotypes be established for terrestrial and oceanic environments in addition to those for epicontinental/marginal marine domains. This TBO - stratotype system, rather than a single kind of stratotypes, would provide a better basis for stratigraphic standardization. On the other hand, we - stratigraphers - are accustomed to observe sedimentary formations in a normal succession. But we are unfamiliar with anomalous successions, such as sedimentary sequences of "melange" type, tectonic slabs and olistostromic units, which are often left for tectonic study alone, without a correct stratigraphic assignment. Thus a special approach by "Non-Smith Stratigraphy" is proposed (though the author is not quite sure of the suitability of its nomenclature).

Keywords: stratotype. TBO-Stratotype System. Non-Smith Stratigraphy

INTRODUCTION

The object of geological research are natural evolutionary processes never experienced by the Recent human being. The factor of time, therefore, serves as key link for the understanding of the geological past. Equivalent approaches can be applied to human civilization history and to Earth history: the reconstruction of a succession of "dynasties" and their mutual correlation in various areas. The main mission of stratigraphy lies exactly in these two aspects: establishment and refinement of the chronostratigrahic scale providing the basis for correlations. For about the last twenty years international stratigraphic studies were focused on the establishment of Global Stratotyes with an attempt to consolidate the geochronological scale in natural stratigraphic sections. This is of great importance for the refinement of the chronostratigraphic scale. In the meantime, both the "International Stratigrahic Guide" and the "Guidelines for the establishment of Global Chronostratigraphic Standards by ICS" (1,2,3,4) also underline the importance of the correlation potential of the stratotypes erected or to be defined. This work goes on, generally-saying, continuously under the guidance of ICS. One of the major difficulties we are dealing with in discussions on GSSPs is the correlatability of biozones between various bioprovinces or realms. Theoretically, transitional successions can help to overcome these problems.

We should, however, feel great concern about the fact that all the stratotypes, both erected or still to be so, are defined or to be placed in epicontinental or marine shelf facies which cannot be correlated directly to other environments. It is well known to all of us that Recent and past epicontinental or shelf seas occupy smaller portions of the Earth's surface than the areas occupied by oceans and land. Not only the first but also the latter can produce relatively continuous stratigraphic records. On the other hand, can stratotypes in shelf or epicontinental marine environments alone satisfy the needs for correlation to other vast realms of the Earth?

NECCESSITY FOR THE ESTABLISHMENT OF A STRATOTYPE SYSTEM
AND OF NON-SMITH STRATIGRAPHY

The environments where sedimentary sequences form can roughly be classified in five categories.

1. Epicontinental or marginal seas, extending to the upper continental slope. In parts of this environment sedimentation is relatively stable and will thus produce a more or less continuous stratigraphic record with abundant biotic remains. Under continental conditions, this kind of successions is often well-preserved through time. All the stratotypes designated so far are located in this paleoenvironment, which, unfortunately, only comprises limited domains and can therefore not be representative for other more extensive paleoenvironments.

2. Land areas. At least since the Late Paleozoic land areas have become more extensive than marginal seas. Although small terrestrial basins are characterized by an unstable sedimentation, large continental depressions, in the contrary, may also produce a continuous stratigraphic record with characteristic biozones uncorrelatable to shallow-marine stratotypes. The Lower Pleistocene Nihewan beds of North China were chosen by the 1948 Geological Congress to be the international standard correlatable to the European terrestrial Villafranchian. However, this attempt to establish non-marine standards failed to be continued.

3. Oceanic realms. They are most extensive and characterized by a stable sedimentation and a continuous stratigraphic record, as demonstrated by the Cenozoic plankton zonation, the best and finest of Phanerozoic biozonations. Although large parts of oceanic sediments have been destroyed by subduction of paleooceanic crust, considerable amounts are still well preserved in association with ancient passive margins and past accretionary wedges. Oceanic sedimentary sequences usually contain conspicuously continuous stratigraphic records and the best zoning fossils. Some characteristic fossils of neritic stratotypes may be absent in oceanic sequences. Nevertheless, common elements, especially planktonic groups like radiolarians and conodonts usually occur in both environments. It can be expected that paleooceanic stratotypes will be a strong tool for a further refinement of the chronostratigraphic scale.

4. Lower Continental Slope with strongly disturbed or reorganized sedimentary sequences. In the case of passive margins, the lower continental slope is characterized by deep sea turbidites or sedimentary mixing. In the case of active margins, deep-sea sediments are usually associated with ophiolites or remnants of oceanic crust, producing so-called " melange" or accretionary wedges, such as the " Franciscan Group" of California, accretionary wedges of the Japan Trench and many other melange complexes preserved on continents. These sequences called melange are due to interactions between continental and oceanic crust. Therefore, the stratigraphic record preserved in them serves as an efficient key to interpret evolutionary histories of continents and oceans. Based on his study of the "Franciscan Group" (5), K. Hsü designated the strata associated with "melanges" as one of the objects of "Non-Smith Stratigraphy".

5. Orogenic domains. Rocks of this domain suffered strongly from geologic disturbances, such as overthrusting, strike-slipping, compression and extension, various kinds of deformation, metamorphism and magmatism. All these processes lead to a reorganization of rock-assemblages and their relationships. Reorganized sequences, though not obeying the rules of traditional stratigraphy, have thus become stratigraphical units due to these secondary formational processes and can reasonably be considered as another object of "Non-Smith Stratigraphy".

In short, to fulfill the mission of stratigraphy imposed on it by geoscience as a whole, there is a need not only of basic stratotypes in shallow-marine environments, but rather of a complete stratotype system (TBO system) including global stratigraphic standards for ancient terrestrial and oceanic environments. Moreover, concerted actions should be taken to develop geohistoric and geodynamic models for the whole Earth. The contribution of Non-Smith stratigraphy to geoscience will lead to a better understanding of ocean-continent interactions through geologic time. Otherwise we, the stratigraphers, would become more and more excluded from the main stream of modern geoscience.

PARTICULARITIES OF STRATOTYPE DEFINITION IN TERRESTRIAL FACIES

In the light of the requirements for stratotype definition, terrestrial sequences have particular properties.

1. They are not characterized by carbonate series but composed mainly of terrigenous clastics. Frequent sandy layers and common intercalations with volcanic materials provide favorable conditions for magnetostratigraphic and radiometric measurements. Although small inland basins with an unstable sedimentation cannot be the subject of a systematic stratigraphic subdivision, large continental depressions with stable and continuous sedimentation can serve as an ideal basis for stratotype definitions.

2. None of the wide-spread nektonic and planktonic taxa correlatable with ammonites, conodonts, nannofossils and foraminifera of marine realms are present in terrestrial successions. However, terrestrial vertebrates, which came into being in the Middle Paleozoic, fill up this gap. Their evolutionary rates were fast enough to be used as leading tools for biostratigraphic zonations, at least good enough for stage or series division or correlation. However, it is a rare case that successive vertebrate zones occur in a single natural section. Candidate sections should herefore be carefully chosen, preferably among sequences deposited in large lake environments with long-term accumulation. Marginal or slope sites of large lakes seem most favourable for the deposition of continuous successions allowing a finer lithologic subdivision and containing vertebrate fossils or those of other terrestrial groups.

3. Terrestrial fossil zonations, too, encounter difficulties due to climatic zones and physical barriers, similar to bioprovince differentiation in marine fossil zonations. However, thanks to the higher ability of temperature-adaptation of land animals, the climatic factor is less important. As to geographic barriers, transitional regions for faunal correlations are relatively easy to find. Fortunately, studies on land-vertebrate dispersal are more advanced than those on marine biotas, as shown by the example of the importance of the Bering land-bridge for the Cenozoic biotic exchange between Eurasia and America, the correlation of Permo-Triassic biotas of Eurasia and Africa, etc. These achievements in terrestrial paleobiogeography speak in favor of the feasibility of stratotype definitions in terrestrial sequences.

PARTICULARITIES OF STRATOTYPE DEFINITION IN PALEOOCEANIC FACIES

1. Oceanic sediments are characterized by low sedimentation rates and high continuity, producing most complete stratigraphic records, typically composed of chemical, biogenic (especially siliceous), and pyroclastic rocks. All these offer good possibilities for chemical, isotopic and magnetic measurements. On the other hand, though they originally covered most of the Earth's surface, they have only a limited preservation potential in natural sections on Recent continents. Nevertheless, a certain amount of them still exists in Recent land outcrops. For example, many natural sections of Early Paleozoic to Cretaceous age occur at some tens of localities in China, constituted of siliceous rocks, pyroclastics and a small amount of terrigenous materials. Many other localities of ancient deep-sea sediments should be preserved all over the world's continents. So far we are not quite sure if all the Phanerozoic stages and series can be recognized in those paleooceanic sediments preserved on the Recent continents. Anyway, most of them can be expected to be present. If so, they will allow a new approach to supplement and calibrate stratotypes

defined in shallow marine facies through establishment of paleoceanic stratotypes.

2. The Phanerozoic deep-sea biota are represented by radiolarians, conodonts, planktonic foraminifera, nannofossils and many other groups which have shown to possess a great potential for biozonation and correlations of world-wide extent. This kind of biozonations appears to be available for the whole Cenozoic and large parts of the Mesozoic and Paleozoic Eras. The accuracy of zonation and correlation based on paleooceanic biota corresponds to stratotype requirements and is by no means inferior in comparison with epicontinental ones.

NON-SMITH STRATIGRAPHY: AN ESSENTIAL CONCEPT

In regions with intense tectonic activities one often meets with geobodies called "groups" or "formations" which, in fact, differ essentially from real groups or formations, as the original associations of sedimentary sequences are strongly deformed or reorganized: stratigraphic units are affected by tectonic displacement, deformation and metamorphism. One should not confuse those different stratigraphical units as this unfortunately often happens in practice. Non-Smith stratigraphy, proposed by K. Hsü in his lectures given in China during the last years, aims particularly at distinguishing those different kinds of stratigraphical units. The present author wrote two articles to introduce the concept to geologists in Chinese publications (6,7). Here three cases must be taken into account.

1. Stratigraphical units in form of "melanges", formed at the bottom of continental slopes adjacent to deep-sea trenchs. In structural geology they correspond to "acccretionary wedges" (or prisms). The matrix usually consist of deep-sea sediments with enclosed exotic blocks, such as dismembered ophiolites - ultramafics, basalts, and, sometimes, olistoliths derived from upper parts of the continental slope. The study by K. Hsü on the "Franciscan Group" of California presented a classical example of this aspect of "Non-Smith Stratigraphy" (7). So far no mature formal terminology has been defined. Would it be appropriate to call them "N-Groups" or "N-Formations"? If so, another problem are the letter symbols to be used in geological maps. The author's proposal is to indicate ages according to the age of the matrix and to place age symbols for enclosed exotic bodies in brackets behind. For example,

"x x x N- Group (or N-formation): J3(C, T)" would imply that the age of the matrix of that unit is Upper Jurassic and its exotic components include Carboniferous and Triassic. This case occurs repeatedly in orogenic domains of China and other countries where mapping practice is often puzzled by these units.

2. Stratigraphical units composed of rock-slabs or overthrusted bodies (nappes). A number of nappes or slabs had been distinguished in the Alps long ago. K. Hsü recognized as such a suite of metamorphosed slabs, previously called "Group" and "Formations" and placed in a progressive stratigraphical column in Shandong Penisula of China. Through this revision, the geologic understanding of the related regions has largely changed.

Despite the facts mentioned above, the slabs themselves, after all, form stratigraphical units. Could they be called "x x x S-Group" (or S- Formation) with their deformation age symbols ahead and formation age symbol of their original rocks (if known) behind? For example,

"x x x S-Group (or S-Formation): J3(Pt3, D)" would imply that the age of deformation is supposed to be Upper Jurassic and that the original rocks were Late Proterozoic and Devonian.

3. Stratigraphical units with synsedimentary olistoliths. Although they may be formed in various environments, they are mostly confined to continental slopes where exotic components range from the size of a sand grain up to hill-like blocks. These units are wide-spread in many orogenic belts and their ages are often indicated taking into account only that of the exotic blocks. The way of their correct indication in maps seems similar to the first two types. For example,

"x x x O-Group (or O-Formation): J3(D, T)" would imply that this unit is of Upper Jurassic age but includes Devonian and Triassic olistoliths which were included in the matrix in the Late Jurassic.

SUMMARY

Moderm geoscience concentrates its main efforts on the obtention of global geodynamic models for the different geologic periods and, therefore, needs stratigraphy in order to broaden its field of study and vision. In this context, the establishment of a stratotype system (TBO system) and the use of Non-Smith stratigraphy are of prime importance. The accomplishment of these steps would help stratigraphy to jump again to the forefront of geoscience where it was situated for more than one century.

REFERENCES

1. H.D. Hedberg (ed.). International stratigraphic guide - a guide to stratigraphic classification, terminology and procedure, 1st ed., *John Wiley and Sons*, 200pp. (1976).
2. A. Salvador (ed.). International stratigraphic guide - a guide to stratigraphic classification, terminology and procedure, 2nd ed., *IUGS & Geol. Soc. Amer.*, 214pp. (1994).
3. J.W. Cowie. Guidelines for boundary stratotypes, *Episodes, 9,* 78-82 (1986).
4. J. Remane, M.G. Bassett, J.W. Cowie, K.H. Gohrbandt, H.R. Lane, O. Michelsen and Wang Naiwen Revised guidelines for the establishment of global chronostratigraphic standards by the International
    Commission on Stratigraphy (ICS), *Episodes*, 19:3, 77-81 (1996).
5. K.J. Hsü Principles of Melanges and their bearing on the Franciscan-Knoxville paradox, *Geol. Soc. Amer. Bull., 79,* 1063-1074 (1968).

6. Wang Naiwen, Guo Xianpu and Liu Yu. A brief introduction to Non-Smith Stratigraphy, *Geol. Rev.*, 40:5, 482 and 394 (in chinese) (1994).
7. Wang Naiwen. Non-Smith Stratigraphy, *Science & China*, 3, 24 (in chinese) (1995).

Proc 30th Int'l. Geol. Congr., Vol. 11 , pp. 29-37
Wang Naiwen and J Remane (Eds)
© VSP 1997

Sequencing, Scaling and Correlation of Stratigraphic Events

FREDERIK P. AGTERBERG
Geological Survey of Canada, 601 Booth Street, Ottawa, K1A 0E0, Canada

FELIX M. GRADSTEIN
Saga Petroleum a.s., Postboks 490, N-1301 Sandvika, Norway

Abstract

The RASC computer program for ranking and scaling of stratigraphic events was developed between 1978 and 1985. It has been available unaltered as version 12 since 1985, and new methodology justified an update commenced in 1994. This paper discusses revised method of scaling stratigraphic events. New results are compared with previous results for a mid-Cretaceous optimum sequence of 87 dinoflagellate, foraminiferal and a log event in 29 wells, northern North Sea and offshore mid-Norway. The improved method (RASC, version 15) yields better definition of clusters of biostratigraphic events, and more sharply defined intercluster distances (breaks) coinciding with hiatuses that can be interpreted in terms of northwestern European sequence stratigraphy. In general, the new scaling procedure is more robust to diminished frequencies of co-occurrences of fossil events in the same wells.

Keywords: Scaling, Correlation, Optimum Sequence, Stratigraphic Events, Cretaceous, North Sea

INTRODUCTION

Expressing uncertainty in biostratigraphic correlation is hampered by a traditionally deterministic approach to zonation and its correlation in stratigraphic sections, especially exploration wells. With this in mind, we set out in 1994 to device a new version of RASC, our computer program for ranking and scaling of stratigraphic events. Based on a probabilistic approach, this program has been available unaltered as version 12 since 1985.

The RASC-12 procedure previously used for scaling has been explained in detail in documentation of FORTRAN computer programs [3, 4], and in other publications or manuals [7, 2: section 6.3]. Extensive applications of RASC during the past 15 years have led to identification of a number of weak points in scaling which could be improved. RASC-15 is based on a revised scaling method which is more robust than RASC-12, as will be discussed in this paper.

In practice, the distribution of fossil events across a number of stratigraphic sections can be highly irregular resulting in small frequencies of individual events and frequencies of pairs of events co-occurring in the same sections. The frequency distribution of microfossil taxa in sections is positively skewed: in general, most taxa occur in one or a few sections only, and relatively few taxa occur in many sections. In RASC, two threshold parameters are set for scaling: (1) k_c for minimum number of sections in which an event should occur, and (2) $m_c < k_c$ for minimum number of sections in which a pair of events should occur before it is used for statistical analysis.

Even when k_c is set relatively high, the probability that two taxa co-occur may remain very small. One possible explanation of this is that many taxa existed within restricted geographic domains occupying parts of the entire study area. If the two domains of two taxa had little overlap, very few sections would contain both taxa. Geographic distribution patterns of fossil events can be studied separately by means of a new computer program (TRACE, in preparation) for traceability, in which the logistic model is used for estimating probability of occurrence of a fossil event at any sampling point including the locations of wells within the study area.

RASC works in the direction of the axis of time. All observed superpositional relations between events (event above/below another event, or pair of coeval events) are projected onto a single axis for relative time. Fossil events are scaled along this relative time axis by performing calculations involving the relative frequencies of the superpositional relations. (An event occurring "above" another event in a section is scored as 1; "below" as 0; and "coeval" as 0.5.) Suppose that two events co-occur in r sections, and that one of the two events has total score of s with respect to the other one, then the relative frequency of this superpositional relation is $p = s/r$. For scaling each p-value is converted into a z-value using the standard normal distribution (probit transformation).

Unless m_c is set equal to a small value, scaling may not be possible for a given data set. For this reason, it is desirable to use a scaling method which is as robust as possible, even when m_c is small. The RASC-12 scaling procedure was as follows: for n events, the maximum number of superpositional relations that can be used to estimate the distance between two successive events (say i and j) is $n^* = n-1$. Each usable pair of events consisting of i or j paired with a third event (say k) results in a value $x_{ij} = z_{ik} - z_{jk}$ which is one of the maximally n^* x_{ij}-values (indirect distance estimates) used to calculate a weighted average value to estimate the interevent distance between i and j along the RASC scale. It is noted that $i = j$ is a special case resulting in a specific x_{ij}-value (direct distance estimate). The weights assigned to the x_{ij}-values depend on the frequencies of all superpositional relations between the three events (i, j and k).

In practice, n^* is much smaller than its maximum value $(= n-1)$ for the following two reasons: (1) the total number of pairs of events involving i or j is reduced by one for each value x_{ij} that cannot be computed, because one or both superpositional relations with k are missing; (2) if i and j occur stratigraphically above (or below) k in all sections where the two events co-occur with k, the resulting two frequencies for superpositional relations are both equal to 1 (or 0), and again x_{ij} cannot be computed. The following tabulation, which is schematical, illustrates which pairs of events can be used and those that can not be used to compute $x_{ij} = z_{ik} - z_{jk}$:

$k=$ 16 17 18 19 20 21 22 23 24 25 26 27 28 29 30 31 32 33 34 35

$i=11$ y y y n n a a a y a y a y a a a a a a a
$j=12$ n y n y n n a y y y a a a a a a a a a a

In this partial table, $x_{ij} = z_{ik} - z_{jk}$ is to be calculated for successive events with $i=11$ and $j=12$, using other events with k going from 16 to 35 only. The first row in the matrix shows possible occurrrences of z_{ik}-values (and the second row z_{jk}-values) as y (yes: $r_{ik} \geq m_c$ and $p_{ik} < 1$), n (no: $r_{ik} < m_c$) or a ($r_{ik} \geq m_c$ and $p_{ik} = 1$). An x_{ij}-value can only be calculated for columns with two y's or y with a (six usable pairs in total).

Special consideration should be given to the latter case (y with a) that only one of a pair of relative frequencies is equal to 1 (or 0). This situation can occur frequently and a large amount of information might be lost if it would be systematically ignored. Neither one of these values (1 or 0) can be readily transformed into a value (z-value) that would be usable for estimating x_{ij} along the RASC scale.

In RASC-12, relative frequencies of 1 and 0 were transformed into AAA and -AAA, respectively. In most applications, AAA ($=$a in the preceding schematic tabulation) was set equal to 1.645 representing the z-value (probit transformation) of a relative frequency of 19 over 20 (or 9.5 over 10). In RASC-15, AAA is replaced by the z-value of $(r-0.5)/r$ which is equal to 1.645 only when $r=10$. With the new, flexible AAA value, the observed "certainty" implied by $p=1$ is replaced by the uncertainty assumption that the two events are coeval in one of the r sections. It is noted that in RASC-12, weight (depending on AAA) was given to pairs of AAA values (a with a in schematic table) for k relatively close to i and j. No such weight is assigned in RASC-15.

Although the preceding revision of scaling rules is relatively simple, it can result in major improvements of the results as will be illustrated on real data in the second part of this paper. The following three, less important, changes were also made in RASC-15:

(1) For small n^* (i.e., relatively few x_{ij}-values), the final interevent distance can be erratic. Although its standard deviation is also estimated in RASC, estimates directly based on the x_{ij}-values are biased [too small, see 2: Section 7.5]. Moreover, even if this source of bias would be small, standard deviations for small n^* are less precise than mean values estimated from the same data. For this reason, RASC-15 has two new rules for setting lower limits on n^*: (a) $n^* \geq 10/m_c$, and (b) total weight $W > 3.5$. Because n^* and m_c are integers, n^* should be at least 4 when m_c is equal to 3 (a frequently used threshold value), at least 5 when $m_c=2$, at least 10 when $m_c=1$, at least 3 when $m_c=4$, etc. The reason that n^* should be larger for smaller m_c is that the x_{ij}-values in the sample then are less precise so that it becomes desirable to use larger samples only. The final interevent distance is a weighted average of the x_{ij}-values in which a weighted sum is divided by the total weight W. Rule (b) sets a lower limit of 3.5 on W. Usually, this rule has little effect. It is an extra safeguard to prevent the use of relatively imprecise x_{ij}-values.

32

(2) Events observed in fewer than k_c sections are not used for computing interevent distances along the RASC scale. The unique (rare) event option allows up to 20 events occurring in fewer than k_c sections to be inserted into the final scaled optimum sequence. In the dendrogram of RASC-12, unique events are plotted in the same way as other events. In RASC-15 dendrograms, the interevent distances involving rare events are not plotted, because they are relatively imprecise. However, the unique events are shown in the scaled optimum sequence to aid in the process of definition of clusters of events.

(3) When RASC is run on data, many relative frequencies (p) are being transformed into z-values, and vice versa. In RASC-12, approximation formulae [1] were used for these transformations: one equation [1: 26.2.23] to derive z from p, and another one [1: 26.2.17] to derive p from z. The first of these two equations has limited precision affecting the interevent distances (usually in the fourth digit after the decimal point). This resulted in a slight drift of the interevent distances when the final reordering option was used. In RASC-15 occurrence of this numerical precison error has been eliminated.

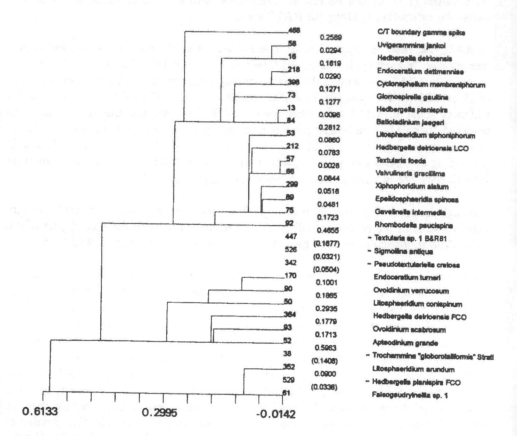

Figure 1. Albian-Cenomanian Scaled Optimum Sequence, offshore mid-Norway. The interevent distances are plotted on the scale to the left. Unique events are preceded by double asterisks. Part of RASC-15 solution with $k_c=6$ and $m_c=3$.

Table 1. Numerical RASC-15 output for example of Fig. 1 (continued on next page)

RANK	FOSSIL PAIRS	INTEREVENT DISTANCE	CUMULATIVE DISTANCE	SUM DIFF. Z-VALUES	SAMPLE SIZE	WEIGHT	S.D.
1	4- 98	0.4266	0.4266	2.6162	6	6.9	0.1264
2	99-155	0.0000	0.4266	0.5646	1	0.9	xxxxxx
3	155- 30	0.2705	0.6961	3.3467	8	12.4	0.0554
4	30- 1	0.0075	0.7036	0.1423	12	19.0	0.1335
5	1- 2	0.0135	0.7171	0.2489	12	18.4	0.0871
6	2- 34	0.1732	0.8903	2.2900	7	13.2	0.1400
7	34-277	0.1420	1.0323	2.2174	9	15.6	0.1764
8	277-186	0.1951	1.2274	4.9166	14	25.2	0.0913
9	186- 29	0.0798	1.3072	2.0062	16	25.2	0.0852
10	29-135	0.2835	1.5907	6.7792	20	23.9	0.1106
11	135-163	0.2741	1.8648	2.5257	9	9.2	0.2154
12	163-234	0.0000	1.8648	-0.2379	1	1.0	xxxxxx
13	234-137	0.0000	1.8648	-0.5545	3	3.2	0.1568
14	137- 70	0.0090	1.8738	0.0621	8	6.9	0.2435
15	70-235	0.2341	2.1080	1.5016	8	8.4	0.2210
16	235- 37	0.3954	2.5034	4.0988	13	10.4	0.1480
17	37- 48	0.0036	2.5071	0.0621	17	14.3	0.1249
18	48- 66	0.1558	2.6629	1.0886	6	7.0	0.1919
19	66- 44	0.0000	2.6629	0.8216	3	2.7	0.0750
20	44- 36	0.3426	3.0054	3.8229	9	11.2	0.1497
21	36- 42	0.2170	3.2225	5.2704	28	24.3	0.1049
22	42-386	0.0355	3.2580	0.4942	18	13.9	0.1404
23	386- 45	0.0506	3.3085	0.5592	10	11.1	0.1500
24	45-290	0.1748	3.4833	2.0976	12	12.0	0.2411
25	290-184	0.2683	3.7517	1.9375	9	7.2	0.3294
26	184-190	0.0556	3.8073	0.8526	10	15.3	0.1860
27	190-276	0.0663	3.8736	1.3108	13	19.8	0.1416
28	276-194	0.4104	4.2840	5.0159	13	12.2	0.1881
29	194-255	0.0008	4.2848	0.0129	14	15.3	0.1146
30	255-370	0.3389	4.6237	4.6825	15	13.8	0.1723
31	370-368	0.2447	4.8684	3.3139	12	13.5	0.1582
32	368-284	0.5069	5.3753	5.2019	10	10.3	0.2152
33	284-330	0.0771	5.4524	0.6328	9	9.2	0.4086
34	330-195	0.1668	5.6192	0.9260	8	5.6	0.1709
35	195-326	0.3650	5.9843	2.3186	11	8.4	0.3140
36	326-369	0.0920	6.0762	0.9173	16	10.0	0.2636
37	369-238	0.1287	6.2049	2.5380	19	19.7	0.1013
38	238-288	0.0633	6.2683	1.0063	15	15.9	0.1374
39	288-205	0.1945	6.4628	2.2777	9	11.7	0.1992
40	205- 82	0.8809	7.1437	11.1935	16	16.4	0.1474
41	82- 27	0.0823	7.2259	0.3551	4	4.3	0.4570
42	27- 26	0.1826	7.4086	0.9150	6	5.0	0.2911
43	26- 63	0.3835	7.7921	2.5473	10	6.6	0.1707
44	63-272	0.1296	7.9216	3.0779	26	23.8	0.1153
45	272- 81	0.1058	8.0274	3.4221	26	32.3	0.1213
46	81-468	0.1358	8.1632	3.8923	22	28.7	0.1267
47	468- 58	0.2589	8.4221	3.5488	14	13.7	0.1542
48	58- 16	0.0294	8.4514	0.4181	13	14.2	0.1147
49	16-218	0.1619	8.6134	4.0562	21	25.0	0.1081
50	218-398	0.0290	8.6424	0.6659	18	23.0	0.1350
51	398- 73	0.1271	8.7695	1.8708	16	14.7	0.1638
52	73- 13	0.1277	8.8972	2.4620	19	19.3	0.1306
53	13- 84	0.0096	8.9068	0.2695	23	26.1	0.1014
54	84- 53	0.2612	9.1680	8.5420	23	30.4	0.1189
55	53-212	0.0860	9.2740	2.8297	24	32.9	0.0863
56	212- 57	0.0783	9.3522	1.8226	19	23.3	0.1520
57	57- 66	0.0028	9.3550	0.0436	12	15.8	0.1959
58	66-299	0.0844	9.4394	1.8245	16	21.6	0.1693
59	299- 89	0.0516	9.4909	1.5414	22	29.9	0.1149
60	89- 75	0.0481	9.5390	0.7505	15	15.6	0.1848
61	75- 92	0.1723	9.7114	2.1652	14	12.6	0.1658
62	92-170	0.4655	10.1768	7.7173	14	16.6	0.0993
63	170- 90	0.1001	10.2769	1.5553	15	15.5	0.1449
64	90- 50	0.1865	10.4634	2.4126	19	12.9	0.1266
65	50-364	0.2935	10.7569	3.1216	15	10.6	0.1575
66	364- 83	0.1779	10.9347	2.0675	12	11.7	0.1443
67	83- 52	0.1713	11.1061	2.7207	14	15.9	0.0990
68	52-352	0.5963	11.7023	7.0979	15	11.9	0.1416
69	352- 81	0.0900	11.7923	0.3396	5	3.8	0.2479
70	81- 96	0.0000	11.7923	1.4861	2	1.5	0.4307

RANK	FOSSIL PAIRS	INTEREVENT DISTANCE	CUMULATIVE DISTANCE	SUM DIFF. Z-VALUES	SAMPLE SIZE	WEIGHT	S.D.
71	68-108	0.8000	11.7823	0.7888	3	3.2	0.1744
72	108-109	0.4506	12.2429	2.8090	5	5.8	0.3558
73	109-103	0.3282	12.5681	1.9817	5	5.7	0.3441
74	103-115	0.1301	12.8982	0.8630	8	6.9	0.2831
75	115-116	0.0768	12.7748	0.5866	6	7.7	0.1723
76	116-107	0.0889	12.8637	0.7661	7	8.6	0.1661
77	107- 59	0.1009	12.9646	0.3936	5	3.9	0.2656
78	59-111	0.0000	12.9646	1.1863	4	2.8	0.2948

COMPARISON OF RASC-15 AND RASC-12 FOR MID-CRETACEOUS MICROFOSSIL EVENTS, NORTH SEA

New results generated with RASC-14 (predecessor of RASC-15) for a mid-Cretaceous optimum sequence of dinoflagellate, foraminiferal and a log event in 29 wells, offshore mid-Norway, have recently been presented [6]. These include optimum sequence of events obtained by ranking followed by variance analysis based on deviations from lines of correlation in the wells. This allows identification, and well to well correlation, of relatively high resolution events to each track best the same stratigraphic level, providing the means to unravel complex sand reservoir stratigraphy. The paper [6] also contains a scaled optimum sequence with definition of RASC clusters for a mid-Cretaceous zonation with interpretation in terms of northwestern European sequence stratigraphy.

The scaling of this data set is used here for comparison of RASC-15 with RASC-12. Fig. 1 shows the scaled optimum sequence in the form of a dendrogram for part of the data set. From the top at event 468 (Cenomanian/Turonian gamma spike) downward there are 3 clusters (58-84: jaegeri/cenomanica RASC zone; 53-92: delrioensis LCO (Lowest Common Occurrence)/siphoniphorum zone; and 170-52: scabrosum zone). The 3 events at the bottom belong to the Falsogaudryinella sp. 1 zone. This late Albian-Turonian zonation with $k_c=6$ and $m_c=3$ contains relatively big breaks at events 92 and 52. These breaks coincide with latest Albian and mid Cenomanian hiatusses, respectively, well-known from NW European sequence stratigraphy. Table 1 shows RASC-15 numerical output pertaining to the computation of the interevent distances plotted in Fig. 1. Most sample sizes exceed 10. As explained in the preceding section, interevent distances based on fewer than 4 x_y-values are automatically set equal to 0 (6 cases in Table 1). Note that the last interevent distance (59-111) was zeroed because its total weight is less than 3.5. The last column contains standard deviations (S.D.) of the x_y-values.

Fig. 2 shows dendrograms obtained using RASC-12 with $k_c=6$ and $m_c=3$. First, AAA was set equal to 1.645 (left side). Next it was set equal to 0.842 (right side) which corresponds to a frequency of 1 over 5 (instead of 1 over 20 for 1.645). The main difference between the two dendrograms of Fig. 2 is that the interevent distances in the diagram on the right are much less (approximately 3 times smaller). This type of dependence of RASC-12 on AAA was previously shown to exist in other data sets as well [5]. The order of the events and the shapes of the clusters show similarity with those in the dendrogram of Fig. 1 which is completely independent of choice of AAA. Results similar to Fig. 2 but obtained by RASC-12 with $k_c=6$ and $m_c=2$ are shown in Fig. 3.

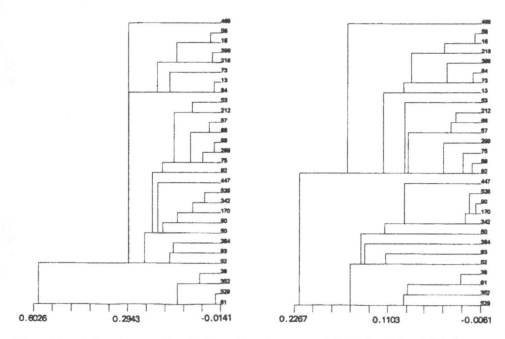

Figure 2. RASC-12 solutions with $k_c=6$ and $m_c=3$. Dendrogram on the left is for AAA = 1.645; the one on the right for AAA = 0.842. Contrary to Fig. 1, RASC-12 results depend on the value selected for AAA.

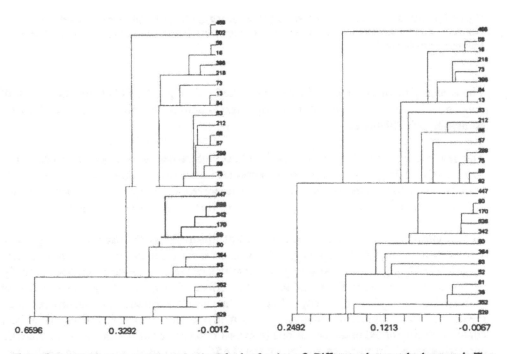

Figure 3. RASC-12 solutions as shown in Fig. 2 for $k_c=6$ and $m_c=2$. Differences between dendrograms in Figs. 2 and 3 are relatively small indicating that change in AAA had more effect than change in m_t.

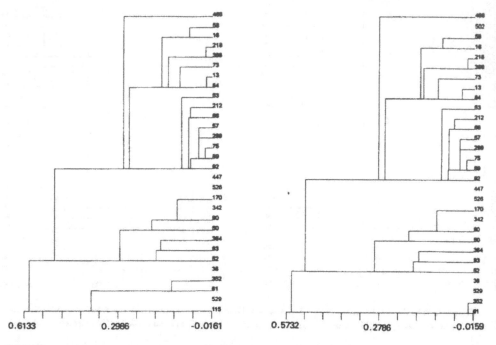

Figure 4. RASC-15 solutions with $k_c=6$, $m_c=2$ (left side), and $k_c=6$, $m_c=1$ (right side). Both dendrograms (as well as the one for $m_c=4$ which is not shown here) closely resemble Fig. 1 indicating robustness of RASC-15 with respect to changes in m_c.

These examples illustrate that, in general, RASC-15 (Fig. 1) yields better definition of clusters of biostratigraphic events and more sharply defined intercluster distances (breaks) coinciding with hiatusses.

The new scaling method is also relatively robust with respect to changes in choice of m_c. The pattern of Fig. 1 was essentially reproduced after setting m_c equal to 2, 1 and 4, respectively, although the sample sizes for the interevent distances then are very different. This robustness is illustrated in Fig. 4 for $m_c=2$ (left side) and $m_c=1$ (right side).

An advantage of the new scaling procedure not illustrated by the example of Figs. 1 to 4 is as follows. In RASC-12 a single AAA value is used for the entire scaled optimum sequence including events near its top and base where sample size for x_{ij}-values gradually decreases to zero. Near these endpoints, the effect of relatively large AAA value is increased. As a result of these edge effects, events near the top and base usually received greater interevent distances than most of the scaled optimum sequence. These edge effects could be distinguished in the deviations (between observed positions in wells and lines of correlation) used for variance analysis when RASC-12 was used. In variance analysis of RASC-15 deviations, this type of bias has been greatly reduced.

CONCLUDING REMARKS

RASC-15 differs from RASC-12 in many features other than scaling. On the one hand, ranking has been improved, primarily by equipping it with options previously restricted to scaling. On the other hand, in RASC-15 both ranking and scaling are followed by variance analysis using deviations of events from lines of correlation fitted to the observed positions of events in the sections. Finally, the probable positions of the events along the RASC scale, along with their error bars, can be projected back onto the depth scale for each section by means of the separate computer program CASC (for correlation and standard-error calculation) which operates on RASC-15 output.

REFERENCES

1. M. Abradowitz and I.A. Stegun (Editors). *Handbook of Mathematical Functions. Applied Math. Ser.*, 55. National Bureau of Standards, Washington, D.C. (1965).
2. F.P. Agterberg. *Automated Stratigraphic Correlation*. Elsevier Amsterdam (1990).
3. F.P.Agterberg and L.D. Nel. Algorithms for scaling of stratigraphic events. *Computers & Geosciences* 8, 163-189 (1982).
4. F.P. Agterberg, F.M. Gradstein, L.D. Lew, M. Heller, W.S. Gradstein, M.A. D'Iorio, D. Gillis and Z. Huang. Program RASC (Ranking and Scaling) version 12. *Comm. Quantitative Stratigraphy*, Bedford Institute of Oceanography, Dartmouth, N.S., Canada (1989).
5. M.A. D'Iorio. Quantitative biostratigraphic analysis of the Cenozoic of the Labrador Shelf and Grand Banks. *Unpublished Ph.D. thesis*, University of Ottawa (1988).
6. F.M. Gradstein and F.P. Agterberg. Uncertainty in stratigraphic correlation. *Proceedings, 'High Resolution Sequence Stratigraphy' Norwegian Petroleum Soc. Conference*, Stavanger, Norway, November 1995. Elsevier, Amsterdam (in press).
7. F.M. Gradstein, F.P. Agterberg, J.C. Brower and W. Schwarzacher. *Quantitative Stratigraphy*. UNESCO, Paris and Reidel, Dordrecht (1985).

Proc 30 *Int'l. Geol. Congr.*, Vol. 11 , pp. 39-43
Wang Naiwen and J. Remane (Eds)
Sedimentation gaps in cyclic sequences.

W.Schwarzacher, School of Geosciences,
Queen's University, Belfast, U. K.

Abstract
A simple model of cyclic sedimentation with randomly distributed gaps is examined. Such incomplete sections will be changed depending on the gap size and the gap frequency. Large gaps or hiatuses can be recognised by differencing and this is illustrated by an example from the Pliocene Trubi marls of Sicily.

Introduction
Defining sedimentation gaps is best done in a negative way by defining a complete (gap free) stratigraphic section. A stratigraphic section is complete when each point on a time scale has a distinct corresponding point on the stratigraphic scale, which is represented by the thickness of the deposited sediment. Sections with gaps always have a lower number of corresponding points than complete sections and they are therefore shorter (see figure 1).

It is clear from the concept of stratigraphic completeness which is scale dependent, that gaps are also scale dependent. Gaps which are below the stratigraphic resolution cannot be recognised.

Cyclic sequences are particularly interesting in the study of gaps because the sedimentary cycle provides a scale which can be used as a reference in examining the distribution of gaps. If the stratigraphic resolution is of the same order as the duration of the cycles, then the problem of sedimentation gaps is clear-cut. In a complete section, the stratigraphic record must contain exactly the same number of cycles as the time sequence. If a gap removes a complete cycle, then the gaps can only be detected by comparing parallel correlated sections. In this case it is not possible to determine which cycle has been removed. This of course assumes ideal cyclicity, in which all the cycles are identical.

The more interesting problem is the recognition of gaps which remove only part of a cycle. Again, the gaps will lead to a shortening of the section but the frequency of the cycles (which includes incomplete cycles) remains the same as the frequency of the time signal.

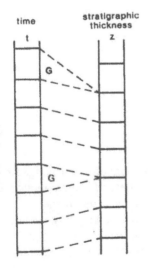

time   stratigraphic
               thickness

Fig.1 Illustrating the definition of gaps
G represents gaps.

A model of incomplete cyclic sedimentation.
To examine the effect of gaps which are smaller than the wavelength of cycles, we investigate the following simple model. We assume that a cyclic signal is represented by a sine wave and $n(t)$ points on the time scale, $t = 0, 1, 2, \ldots n(t)$. The series contains f cycles of wave length λ. If the sedimentary record is complete, it will contain $n(z)$

corresponding points on the stratigraphic scale. We assume that the stratigraphic record contains a number of gaps which occur at random with the probability p. The size of the gaps g, is assumed to be constant and it is measured in units of the wave length. A section of n points will be shortened to length l(z) and it will contain λpg gaps.

$$l(z) = \frac{n(t)}{1 + pg\lambda}$$

Figure 2 gives two examples of simulated records, both representing a shortening of 50 %, with small gaps (g = 0.1, p = 0.5) for case A and relatively large gaps (g = 0.2, p = 0.25) for case B.

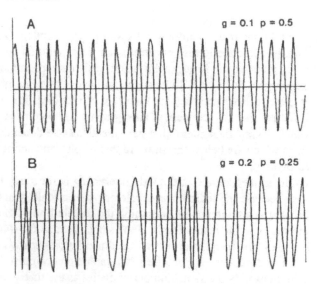

Fig.2 Two examples of sine waves containing random gaps.

The amount of shortening of a section is determined by the product pg. Clearly, the same amount of shortening of a section can be achieved by a wide variety of combinations of gap sizes and frequencies. It is important to recognise that with the same overall shortening, very different records can develop. If the gap size is relatively large, a few gaps can remove complete cycles and reduce the total number of cycles. Furthermore, large gaps produce abrupt phase shifts and power spectra from such records deteriorate. When the gaps are small and frequent, considerable shortening of sections may occur without changing the spectrum.

Figure 3 shows the deterioration of the spectra which can be clearly seen by comparing the power spectra of the two examples from figure 2.

In most cases, shortening of the section will reduce the number of recorded cycles and therefore will lead to a shift of the frequency maxima towards the lower end. The actual amount of the frequency shift is difficult to evaluate because the truncation of the cycles alters their shape and it is not obvious how incomplete cycles can be counted.

To obtain some information on how cyclic sections with sedimentation gaps behave, a number of simulations have been performed. Each consisted of 1000 time points forming a

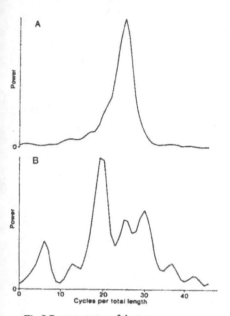

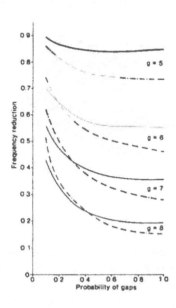

Fig.3 Power spectra of the two
examples in fig.2.

Fig.4 The frequency reduction (y axis) for simulated
sine waves with gaps and frequencies. The solid line
represents 40 cycles per run, the broken line
represents 80 cycles per run.
The x axis gives the probability of gap occurrence.

sine wave with values from -1 to +1. Each series contained f(t) cycles. Gaps of size $g\lambda$
and density p were inserted and after this operation, the remaining cycles were counted. A
cycle boundary was defined as any change from a negative to a positive value. The count
of such boundaries represents the reduced frequency and the ratio $f(z)/f(n)$ which is shown
along the ordinate of figure 4, and represents the signal frequency reduction for various
frequencies, gap sizes and gap densities.

Since the effect of gaps is scale dependent, the presence of gaps will also have different
effects on cycles with different frequencies.The simulations show the decrease of the
frequency ratio $f(z)/f(n)$, with increasing frequency of the signal, increasing size of gaps
and increasing density of gaps. Cycles with high frequencies will be preferentially
eliminated and the shift of high frequency maxima is faster, compared with the longer
wave lengths.

Although it will be difficult in practice to obtain quantitative data on gap sizes and
distributions in real sections, the results of the simulations are relevant when interpreting
sections containing cycles of different orders.

For example in sections recording Milankovitch cycles, one could expect two cycles
(precession and eccentricity) with a wave length ratio of approximately 1 : 5. Randomly
distributed gaps can reduce this 1 : 5 ratio to 1 :4 or even 1 :3. Similar reductions of ratios
can be caused by random fluctuations in rates of sedimentation or by bioturbation.
(Schwarzacher 1993).

The above model can be generalised by allowing variable gap sizes which are controlled
by some statistical parameter. It is also possible to vary the distribution of gaps. In many
real examples, gaps do not occur at random. For example in cycles produced by eustatic
sea level changes, erosion or gap formation is most likely to be restricted to sea level low
stands or during emersion. The resulting shortened records, not surprisingly resemble
cycloids which are a well-known model for cycles with variable sedimentation rates,
including sedimentation still stands and erosion (Schwarzacher 1975). The frequency of

the time record will be preserved if the truncation of the cycles always occurs in the same place, otherwise phase shifts occur and the spectra will deteriorate.

The recognition of large sedimentation gaps or hiatuses.
As previously mentioned, the effect of randomly distributed gaps is identical to randomly fluctuating sedimentation rates and indeed sedimentation rates on a microscale are controlled by the presence of numerous sedimentation gaps. Once again,the definition of a hiatus is a question of scale. A gap or hiatus is generally regarded as being a sufficiently long time interval to remove part of a recognisable segment of stratigraphic evolution. Such a segment may be a part of a cycle or part of any trend in the history of the sediment. The hiatus is recognised as a break in such a trend.The recognition of trends is made difficult by several limitations. The most important of these is the availability of adequate stratigraphic resolution data but also random noise which is caused either by fluctuations in sedimentation rates or indeed by noise in the original time record. If the random element in the original record is so strong that no trend can be recognised, then it is impossible to recognise a hiatus-like gap. Noise can be reduced by various filtering methods and filtering will always be the first step in the search for significant sedimentation gaps. Fortunately, there are a large number of observable sedimentological features which can indicate such breaks in sedimentation.
If trends are present in the record, discontinuities can be found by differentiation of the time series. This is achieved by calculating finite differences of several orders. If the series is well-behaved (without discontinuities) the differences of higher order will decrease and may oscillate around a constant value.
If the series contains a discontinuity, a disturbance is generated which spreads and increases in magnitude with higher differences. If one is dealing with a series of random numbers, no convergence of higher orders occurs and the spread of the differences remains constant through any order. The application of difference analysis often needs considerable experimentation in practice. A borehole in the Pliocene Trubi marls provided the following three types of information
1) Visual inspection of the cut and smoothed core surface.
2) At 5 cm intervals, colour and bioturbation were estimated using a nominal scale.
3) Biostratigraphic data (both planktonic and benthic foraminifera counts) at 20 cm intervals.
All three variables indicate strong cyclic sedimentation with cycle thicknesses of approximately 1 m. This cycle is believed to be precession controlled (Hilgen 1987). Consider first the biostratigraphic data. It is obvious that each cycle is only represented by about 4 to 5 datapoints and the resolution compared with the wavelength of the cycle, is not sufficient to detect any changes within the cycles. The colour and bioturbation data on the other hand, are more detailed but extremely noisy.To obtain good results from these data, they have to be filtered. The degree of filtering is best established by experiment. If filtering is too much, all discontinuities disappear and if it is insufficient, no stable maxima in the differenced series are found. A simple running average repeated three times gave satisfactory results.Differences which are higher than third order differences, remain in the same position and coincide with breaks in sedimentation which have been recognised during the visual inspection of the core (see fig.5). This coincidence with actually observed breaks, confirms that part of the cyclic sequence is missing but it cannot

tell if it is only a part of a cycle or if one or more complete cycles are missing. It is of some interest that the highest difference is due to a break in the core which had not been interpreted correctly and which would have gone unnoticed without difference analysis.

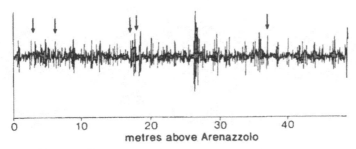

metres above Arenazzolo

Fig.5 Third order differences of colour index values in a core from the Pliocene Trubi marl. Discontinuities which have been identified by inspection of the core, are indicated by arrows. The large maximum is due to a break in the core.

Conclusions.
The study of the interrupted sine wave model has shown that sedimentation gaps can strongly modify the record of a time history. High cycle frequencies are more rapidly changed than the lower frequencies. The size of gaps with respect to the the cycle length is of prime importance. It is likely that detailed sedimentological studies will be more successful than analytical methods, in recognising sedimentation gaps.

References.

Hilgen F.J., 1987. Sedimentary rhythms and high-resolution chronostratigraphic correlations in the Mediteranian Pliocene. Newsl. Stratigr. **17**, 109- 127.
Schwarzacher W. 1975. Sedimentation models and quantitative stratigraphy. Elsevier, Amsterdam,382 pp.
Schwarzacher W. 1993.Cyclostratigraphy and the Milankovitch Theory. Elsevier, Amsterdam. 225 pp.

Proc 30th Int'l. Geol. Congr., Vol. 11 , pp. 45-54
Wang Naiwen and J. Remane (Eds)
VSP 1997

THE APPLICATION OF RASC/CASC METHODS TO QUANTITATIVE BIOSTRATIGRAPHIC CORRELATION OF NEOGENE IN NORTHERN SOUTH CHINA SEA

WANG PING and ZHOU DI

South China Sea Institute of Oceanology, Academia Sinica, Guangzhou 510301, China

Abstract

RAnking and SCaling (RASC) and CorrelAtion and SCaling (CASC) are statistically carefully designed methods performing the computation of optimal biostratigraphic sequence, biozonation, and inter-well chronological correlations. In the Zhujiangkou Basin on the northeastern shelf of the South China Sea, 34 wells with relatively complete fossil records were selected for quantitative biostratigraphic analysis using RASC/CASC methods. 112 fossil events were used (55 calcareous nannofossils, 4 foraminifera, and 10 sporopollen), as they occurred in at least 6 wells, or they were important index events. The optimum sequence and the RASC zonation of these events were obtained. Ten assemblage zones and 15 subzones were divided by large RASC distances. Then the RASC zones are correlated with the global biochronozones, and with lithological divisions and seismic reflectors of the region. A geological time table is constructed. In general the optimum sequence agrees well with global biochronozones. Two exceptions revealed the questionable allocation of the LAD of *Globorotalia limbata* in N21 zone and problems in the correlation of fossil events in the N15 zone. 17 wells were selected for chronological correlation by using CASC method. The isochrons of 5 Ma, 10 Ma, and 24 Ma respectively agree with lower boundaries of Wanshan, Yuehai, and Zhujiang formations, but the 16 Ma isochron is consistently higher than the lithological lower boundary of the Hanjiang Formation. It suggests that the lower boundary of this formation is not the bottom of Mid Miocene, but a lithological boundary. Thus our quantitative analysis lead the long time debate on this boundary to a conclusion. Age-depth curves of well sections were derived from the age-event curve. These made possible a high resolution analysis of Neogene subsidence history of the region.

Keywords: quantitative biostratigraphy, Neogene, northern South China Sea

INTRODUCTION

Methods of RAnking and SCaling (RASC) and CorrelAtion and SCaling (CASC) were introduced to quantitative biostratigraphy in early 80s and underwent several steps of refinement later [6,2,1]. Based on observed sequences of biostratigraphic events from multiple wells, these statistically carefully designed methods enable us to compute the optimal biostratigraphic sequence, biozonation, biostratigraphic time table for the area, and to perform inter-well chronological correlations. The methods have been successfully used in northwest Atlantic margin [6], Grand Banks and Labrador shelf [7], and other places in the world. This paper presents an application of RASC/CASC methods to the Neogene biostratigraphy of the sedimentary basins in northern South China Sea. Results indicate the methods are powerful in manipulating multi-well data to give more subjective and representative biostratigraphic sequence and inter-well chronological correlations. This enables the calculation of sedimentary rate and an analysis of tectonic subsidence with improved resolution

RASC/CASC METHODS

RASC/CASC computations are based on sequences of biostratigraphic events observed from multiple well or/and geological sections in the study region [7]. A biostratigraphic event is a special event of organism development which occurs only once in the time scale, for example, the last (LAD), the first (FAD), and the peak (ACME) appearance of a species. The sequence of events should be unique within the same biological province. Observed sequence of events, however, often deviate from well to well or section to section due to many random factors during sediment deposition, preservation, sampling, as well as fossil examination. The conceptual basis of RASC is that the biostratigraphic sequence and zonation given by statistical averaging of a large number of observations are usually less affected by random error, less subjective than those synthesized by human's brain. Such sequence and zonation are more reliable and may be called optimum.

RASC performs ranking and zonation. Ranking is to find the optimum sequence of the fossil events based on observed sequences. By applying a permutation algorithm to the event occurrence matrix, an optimum sequence may be found whose occurrence is maximum. Thus the optimum sequence is the most likely sequence of the events according to given data set. Zonation is to identify event assemblage zones. This involves the definition of inter-event distance as the measure of the closeness of events in the optimum sequence. The distance between events A and B is defined as the normal deviate with respect to the probability of A over B (P_{AB}) in the sections, Z_{AB}. For example, if P_{AB}=0.5, then Z_{AB}=0, events A and B are coeval in time scale. The larger the P_{AB} deviated from 0.5, the larger is Z_{AB}. When P_{AB}=1, then Z_{AB}=1, event A is always over B. Based on inter-event distances in the optimum sequence, a dendrogram is drawn, from which event assemblage zones may be divided by large distances.

CASC calculate the most likely depths of events in wells. The key in CASC is to estimate ages of the events in the optimum sequence based on their inter-event distances. This is achieved by a cubic spline fitting of the curve of age vs. cumulative inter-event distance, using the index fossils of known age (or other age data) as nodal points. When ages of all events are known, a geological time table for the region may be compiled in conjunction with other geological information. Then inter-well correlation can be achieved even for sections lack of index fossils. Then a high resolution analysis of subsidence history of the region may be performed.

NEOGENE STRATIGRAPHY OF NORTHERN SOUTH CHINA SEA

On the northern shelf of the South China Sea developed several petroliferous Tertiary sedimentary basins, among which the Zhujiangkou (Pearl River Mouth) Basin is the largest one (Fig. 1). By the end of 1996, a total of 23 oil-bearing structures have been found in this basin, 8 oil fields have been in production, and 11 million tons of annual production is anticipated. Now the northern shelf of the South China Sea becomes China's   fourth largest base for oil exploration and production.

For the northern shelf of the South China Sea, Neogene was the period of thermal subsidence after extensive rifting and continental sedimentation in Paleogene. In Late Oligocene the shelf broke up, the South China Sea basin began to form, and the transgression on the northern shelf started. The strata of Late Oligocene and Neogene in

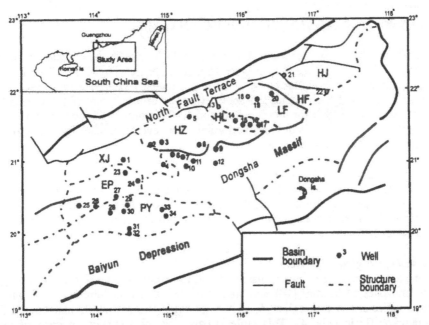

Figure 1. Map of the central and eastern Zhujiangkou Basin, showing major structural elements and localities of the wells used in this study. PY - Panyu Low Uplift; EP - Enping Sag; XJ - Xijiang Sag; HZ - Huizhou Sag; HL - Huilu Low Uplift; LF - Lufeng Sag; HF - Haifeng Uplift; HJ - Hanjiang Sag.

the area consist of entirely marine sediments, mainly mudstone, siltstone, and sand-stone of coastal plain, deltaic, and neritic facies, with less amount of reefal limestone (Fig. 2). The lower section of Neogene contains mainly benthic foraminifera, suggest-ing a relatively shallow water. Since Late Miocene the area became an open sea, as indi-cated by abundant planktonic fora-minifera and calcareous nan-nofossils. Fossil records indi-cate that the area was mostly in the subtropic zone with warm and humid climate, except in Late Oligocene to early Early Miocene when the area was relatively cool and dry. There were three major sedimentary cycles, with trans-gression climax occurring in late Early Miocene, middle Late Miocene, and Early Plio-cene. Sand bodies formed at the beginning of transgress-ions and the ending of regressions, as well as reef buildups formed in shoaling areas are good reservoirs for hydrocarbon, and transgre-ssional mudstones are good caps [5, 11].

AGE (Ma)	SERIES	FORMA-TION	THICKNESS (m)	REFLECTOR	LITHOLOGY	OIL ZONE	SEDIMENTARY FACIES	SEALEVEL CHANGE
	Quaternary	Wan-shan	200	Tn				High Low
5	Pliocene		100 450	T₁			Open shelf	
	U	Yue-hai	200 600					
10		Han-jiang	500 1100	T₂		•	Deltaic	
15	M			T₄		•	Embayment	
20	L	Zhu-jiang	350 750			•	Shelf	
				T₇		¤	Carbonate platform	
25	OLIGO-CENE	Zhu-hai	500 1100			•	Paralic	

Figure 2. Neogene stratigraphic column for the Zhujiangkou Basin (after Chen and Li [5]).

48

QUANTITATIVE BIOSTRATIGRAPHIC ANALYSIS

In order to obtain an optimum biostratigraphic sequence and a better inter-well correlation, RASC/CASC methods were applied to analyze fossil records of Neogene strata in the Zhujiangkou Basin. 34 wells with relatively complete fossil records were selected for the analysis. Among 268 fossil events recorded, a total of 112 events (55 for calcareous nannofossil, 4 for foraminifera, and 10 for sporopollen) were retained, as they occurred in at least 6 wells or were important index events. Except for sporopollen, the events are mainly last appearance.

Optimal Sequence and Zonation

Through RASC computation, the optimum sequence of 112 events was obtained (Fig.3). Ten event assemblage zones (hereafter called RASC zones) were identified with the interzonal distances greater than 0.26, and 15 sub-zones were divided by distances greater than 0.16.

Regional Geological Timetable

According to positions of index fossils, the RASC zones are correlated with the global biochronozones of Martini [10] and Haq et al. [8], and with lithological divisions and seismic reflectors of the region [5, 11]. A geological time table is constructed for the Zhujiangkou Basin (Fig. 4). This table is based on the statistical average of multiple wells and thus more reliable than that constructed based on qualitative synthesis.

Boundaries between RASC zones correspond well with cycles of sealevel changes and seismic reflectors. The most important RASC boundary appears between R1 and R2 with an inter-event distance of 0.48, the highest one in the entire sequence. This corresponds to the Paleogene/Neogene unconformity with a significant change from the continental facies to marine facies. Other major boundaries correspond to significant regressions, including 1) the R2/R3 boundary with a distance of 0.37 and age of about 19 Ma, representing the end of a long regression period; 2) the R5/R6 boundary with a distance of 0.36, corresponding to the T2 reflector and the boundary between the Hanjiang and Yuehai formations, and indicating a big regression at the beginning of Late Miocene; and 3) the R8/R9 boundary in the upper part of the Pliocene Wanshan Formation, corresponding to a large regression with some foraminifera of N21 zone missing [16].

The optimum sequence agrees well with global biochronozones, except the position of *Globorotalia limbata* (event 198). According to Blow [3], this species belongs to the N21 zone, but in our study it appears in the R8 zone together with calcareous nannofossils of NN16 and NN15 zones of middle Pliocene. The chronological allocation of this fossil event in the Zhujiangkou Basin is questionable. Another problem is the correlation of the fossil events in the N15 zone. Blow [3] used the FAD of *Globorotalia acostaensis* to define the top of the N15 zone and correlated it with the NN8 zone, dated as late Mid Miocene. But Hag et al. [8] correlated N15 with NN9 of early Late Miocene. In the Zhujiangkou Basin, LAD of *G. continuosa* (event 184) and *Globoroquadrina ehiscens advena* (event 176) are usually used to replace the FAD of *G. acostaensis*, as FAD is hardly to identify in well sections. By Kennett and Srinivana [9], all the three events belong to N15. In our study, the two LADs are associated with events of NN10 and NN9; *G. continuosa* is located in between NN10 and NN9, closer to NN10 according to inter-event distance. Thus we suggest that the LAD of *G. continuosa* is an event within N16, and the N15 zone should be correlated with the lower portion of NN9.

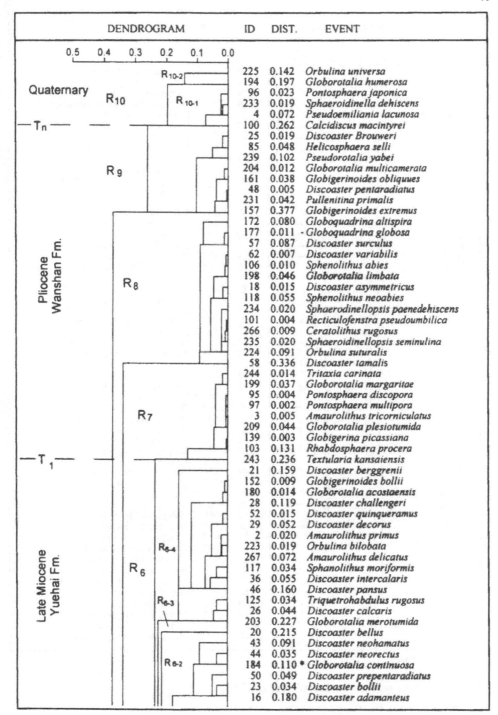

DENDROGRAM	ID	DIST.	EVENT

0.5 0.4 0.3 0.2 0.1 0.0

ID	DIST.	EVENT
225	0.142	*Orbulina universa*
194	0.197	*Globorotalia humerosa*
96	0.023	*Pontosphaera japonica*
233	0.019	*Sphaeroidinella dehiscens*
4	0.072	*Pseudoemiliania lacunosa*
100	0.262	*Calcidiscus macintyrei*
25	0.019	*Discoaster Brouweri*
85	0.048	*Helicosphaera selli*
239	0.102	*Pseudorotalia yabei*
204	0.012	*Globorotalia multicamerata*
161	0.038	*Globigerinoides obliquues*
48	0.005	*Discoaster pentaradiatus*
231	0.042	*Pullenitina primalis*
157	0.377	*Globigerinoides extremus*
172	0.080	*Globoquadrina altispira*
177	0.011	*Globoquadrina globosa*
57	0.087	*Discoaster surculus*
62	0.007	*Discoaster variabilis*
106	0.010	*Sphenolithus abies*
198	0.046	*Globorotalia limbata*
18	0.015	*Discoaster asymmetricus*
118	0.055	*Sphenolithus neoabies*
234	0.020	*Sphaerodinellopsis paenedehiscens*
101	0.004	*Recticulofenstra pseudoumbilica*
266	0.009	*Ceratolithus rugosus*
235	0.020	*Sphaeroidinellopsis seminulina*
224	0.091	*Orbulina suturalis*
58	0.336	*Discoaster tamalis*
244	0.014	*Tritaxia carinata*
199	0.037	*Globorotalia margaritae*
95	0.004	*Pontosphaera discopora*
97	0.002	*Pontosphaera multipora*
3	0.005	*Amaurolithus tricorniculatus*
209	0.044	*Globorotalia plesiotumida*
139	0.003	*Globigerina picassiana*
103	0.131	*Rhabdosphaera procera*
243	0.236	*Textularia kansaiensis*
21	0.159	*Discoaster berggrenii*
152	0.009	*Globigerinoides bollii*
180	0.014	*Globorotalia acostaensis*
28	0.119	*Discoaster challengeri*
52	0.015	*Discoaster quinqueramus*
29	0.052	*Discoaster decorus*
2	0.020	*Amaurolithus primus*
223	0.019	*Orbulina bilobata*
267	0.072	*Amaurolithus delicatus*
117	0.034	*Sphanolithus moriformis*
36	0.055	*Discoaster intercalaris*
46	0.160	*Discoaster pansus*
125	0.034	*Triquetrohabdulus rugosus*
26	0.044	*Discoaster calcaris*
203	0.227	*Globorotalia merotumida*
20	0.215	*Discoaster bellus*
43	0.091	*Discoaster neohamatus*
44	0.035	*Discoaster neorectus*
184	0.110 *	*Globorotalia continuosa*
50	0.049	*Discoaster prepentaradiatus*
23	0.034	*Discoaster bollii*
16	0.180	*Discoaster adamanteus*

Figure 3. Dendrogram of scaled optimum sequence of biostratigraphic events for the Zhujiangkou Basin.
* - unique event; *LAD* - last appearance; *FAD* - first appearance; *DFCA* - last continuous appearance; *ACME* - peak appearance. Unspecified events are all *LAD*.

50

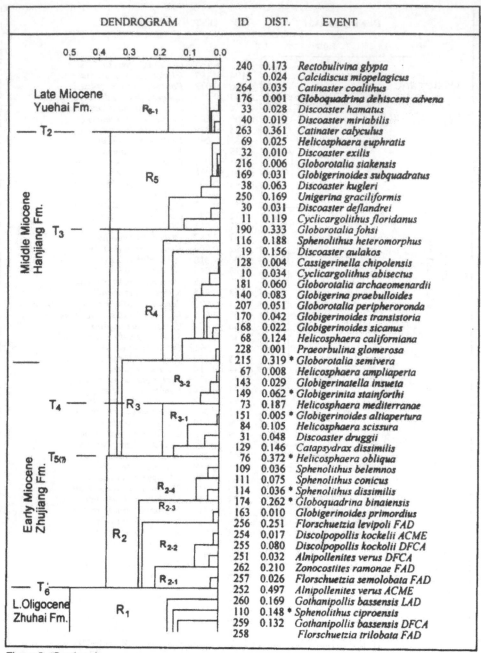

DENDROGRAM	ID	DIST.	EVENT
	240	0.173	*Rectobulivina glypta*
	5	0.024	*Calcidiscus miopelagicus*
	264	0.035	*Catinaster coalithus*
	176	0.001	*Globoquadrina dehiscens advena*
	33	0.028	*Discoaster hamatus*
	40	0.019	*Discoaster miriabilis*
	263	0.361	*Catinater calyculus*
	69	0.025	*Helicosphaera euphratis*
	32	0.010	*Discoaster exilis*
	216	0.006	*Globorotalia siakensis*
	169	0.031	*Globigerinoides subquadratus*
	38	0.063	*Discoaster kugleri*
	250	0.169	*Unigerina graciliformis*
	30	0.031	*Discoaster deflandrei*
	11	0.119	*Cyclicargolithus floridanus*
	190	0.333	*Globorotalia fohsi*
	116	0.188	*Sphenolithus heteromorphus*
	19	0.156	*Discoaster aulakos*
	128	0.004	*Cassigerinella chipolensis*
	10	0.034	*Cyclicargolithus abisectus*
	181	0.060	*Globorotalia archaeomenardii*
	140	0.083	*Globigerina praebulloides*
	207	0.051	*Globorotalia peripheroronda*
	170	0.042	*Globigerinoides transistoria*
	168	0.022	*Globigerinoides sicamus*
	68	0.124	*Helicosphaera californiana*
	228	0.001	*Praeorbulina glomerosa*
	215	0.319 *	*Globorotalia semivera*
	67	0.008	*Helicosphaera ampliaperta*
	143	0.029	*Globigerinatella insueta*
	149	0.062 *	*Globigerinita stainforthi*
	73	0.187	*Helicosphaera mediterranae*
	151	0.005 *	*Globigerinoides altiapertura*
	84	0.105	*Helicosphaera scissura*
	31	0.048	*Discoaster druggii*
	129	0.146	*Catapsydrax dissimilis*
	76	0.372 *	*Helicosphaera obliqua*
	109	0.036	*Sphenolithus belemnos*
	111	0.075	*Sphenolithus conicus*
	114	0.036 *	*Sphenolithus dissimilis*
	174	0.262 *	*Globoquadrina binaiensis*
	163	0.010	*Globigerinoides primordius*
	256	0.251	*Florschuetzia levipoli FAD*
	254	0.017	*Discolpopollis kockelii ACME*
	255	0.080	*Discolpopollis kockolii DFCA*
	251	0.032	*Alnipollenites verus DFCA*
	262	0.210	*Zonocostites ramonae FAD*
	257	0.026	*Florschuetzia semolobata FAD*
	252	0.497	*Alnipollenites verus ACME*
	260	0.169	*Gothanipollis bassensis LAD*
	110	0.148 *	*Sphenolithus ciproensis*
	259	0.132	*Gothanipollis bassensis DFCA*
	258		*Florschuetzia trilobata FAD*

Figure 3. (Continued)

This correlation agrees with that of Haq *et al.* [8].

Inter-well Correlation
The CASC program gives an interpolated age-event curve, from which the age for each event in the optimum sequence may be read out. This greatly facilitates the inter-well

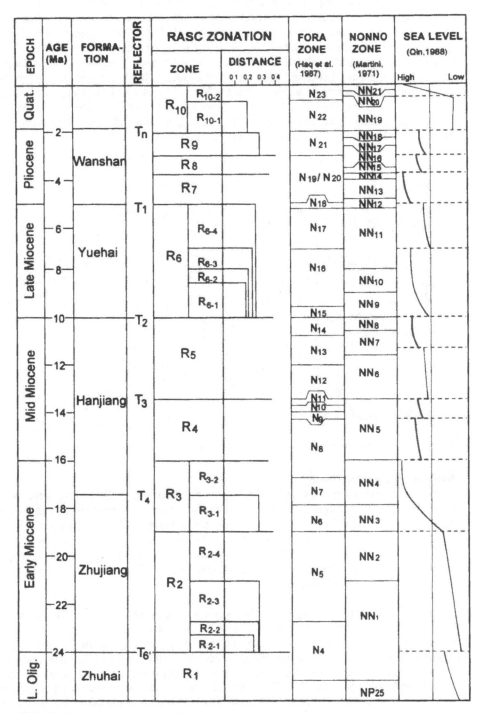

Figure 4. Neogene time table for the Zhujiangkou Basin.

52

correlation, especially for the wells with less or no index fossils. We selected 17 wells for chronological correlation (Fig.5). Results show that isochrons of 5 Ma, 10 Ma, and 24 Ma respectively agree with lower boundaries of Wanshan, Yuehai, and Zhujiang formations, but the 16 Ma isochron is consistently higher than the lithological boundary between Hanjiang and Zhujiang formations). The difference is 56-440 m, in average 215 m, smaller difference occurring only on uplifts. Such a large difference is not possibly caused by random errors. It suggests that the lower boundary of the Hanjiang Formation is not the bottom of Mid Miocene, but a lithological boundary within Lower Miocene. There was a long time discussion on the dating of this lithological boundary. Our probablistic correlation brought this debate to a conclusion.

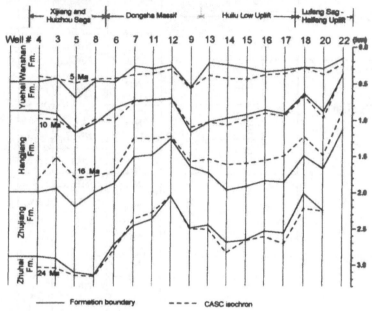

Figure 5. Inter-well correlation for the Zhujiangkou Basin. Numbers on the top are well codes as shown in Fig. 1.

Analysis of subsidence history

From the age-event curve given by CASC computation, age-depth curves were derived for each wells in the region. These constitute the bases for a subsidence analysis with higher resolution by using the BURSUB program of Stam *et al.* [15]. This analysis show that the region was in a gross subsidence of 1-2 km during the Neogene period. The subsidence history is similar for wells within the same depression or uplift, but significantly different between depressions and uplifts. The amplitude of subsidence increased southwestward in general.

Then subsidence rates for representative wells are computed (Fig. 6). The subsidence was fast in early Early Miocene and much slower after then. A sharp decrease in subsidence rate occurred in 19-20 Ma, corresponding to the top of R2 zone. It was suggested that the Philippine Plate obliquely collided with the Eurasian Plate at about 20 Ma, and this caused the sudden change of the spreading direction of the South China Sea from NS to NNW-SSE [13, 4]. Differential subsidence was strong also in other two times, at the end of Mid Miocene and in Pliocene [14]. These were the dates for the first and second phase of the Dongsha Movement [5]. Fig. 7 shows that the first phase was manifested by the significant slow down of the subsidence in Dongsha massif and Xijiang, Huizhou sags; while the second phase was shown most strongly by the uprising of the Panyu Low Uplift.

DISCUSSIONS

The application of RASC/CASC methods to the quantitative stratigraphic correlation of Neogene strata in the Zhujiangkou basin demonstrated the following merits of the methods: 1) It provides a probabilistic summary of biostratigraphic records for the area. In particular, the most likely (optimum) sequence of fossil events and their assemblage zones, which places recorded events, not only index fossils, into united sequence and zonation which fit best the observed data and thus are more subjective and representative for the area. 2) It estimates ages for all the events in the sequence based on their inter-event distances. This greatly facilitates the construction of regional geological time table, and make possible a regional biostratigraphic

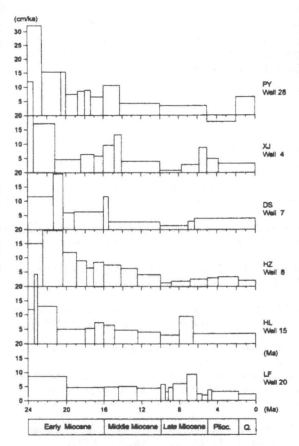

Figure 6. Rates of tectonic subsidence for selected wells from the Zhujiangkou Basin. Numbers on the right side are well codes. Notations of structural divisions are the same as those in Fig.1

Correlation that includes also sections with sparse index fossils, and a regional subsidence analysis with improved resolution and reliability.

The advantages of RASC/CASC methods are apparent when dealing with biostratigraphic data of considerably large number of sections and events, such as in the above-mentioned application. Otherwise their results may be unstable. In another application to a basin in the western portion of the northern South China Sea, out of 158 events recorded, only 78 events occur in 3 or more wells, and only 22 events in 6 or more wells. When a threshold of 3 was used, the optimum sequence became very sensitive to a slight change of input, such as a deletion of one bad record. Using the weighting function in the program usually did not help. A higher threshold will decrease the resolution badly. Although this outcome is anticipated as the probabilistic nature of the methods, some improvement may be done to the algorithm. For example, some constrains may be introduced to the calculation of the cross-over probability of events in order to improve its resistivity to outliers.

54

Acknowledgments

Jointed the study were also Fang Qing and Li Ming-xing. The authors thank Nanhai East and Nanhai West Oil Co., China National Oil Corp. for financing and providing basic data for the study. We are grateful to Agterberg and Gradstein for kindly providing RASC/CASC programs, and to our colleagues for helpful discussions.

REFERENCES

1. Agterberg,F.P. Automated Stratigraphic Correlation, Elsevier, New York (1990).
2. Agterberg,F.P. and Byron,D.N. Micro-RASC system of 12 Fortran 77 microcomputer programs for ranking, scaling and correlation of stratigraphic events. Geological Survey of Canada, Open File Report (1989).
3. Blow,W.H. Late Middle Eocene to recent planktonic foraminiferal biostratigraphy. In *Proceedings of the First International Conference on Planktonic Microfossils*. R.Broennimann and H.H.Renz (Eds.), pp.199-421. E.J.Bill, Leiden (1969).
4. Briais,A., Patriat,P., and Tapponnier,P. Updated interpretation of magnetic anomalies and seafloor spreading stages in the South China Sea: Implications for the Tertiary tectonics of Southeast Asia. *J. Geophys. Res.* 98, 6299-6328 (1993).
5. Chen, Sizhong and Li, Zesong. Review and prospects for the oil and gas exploration and development of Eastern Pearl River Mouth Basin, *China Offshore Oil and Gas (Geol.)* 6(2), 21-30 (in Chinese, with English abstract, 1987).
6. Gradstein,F.M. and Agterberg,F.P. Models of Cenozoic foraminiferal stratigraphy, northwestern Atlantic margin. In: *Quantitative Stratigraphic Correlation*. J.M.Cubitt and R.A.Reyment (Eds.), pp.119-173. John Wiley and Sons, Chichester (1982).
7. Gradstein,F.M., and Agterberg,F.P. Quantitative correlation in exploration micropaleontology. In: *Quantitative Stratigraphy*. F.M.Gradstein et al. (Eds.). pp. 309-357, UNESCO, Paris (1985).
8. Haq, B. U., Hardenbol, J., and Vail, P. R. Mesozoic and Cenozoic chronostratigraphy and cycle of sealevel change, in *Sea Level Change: An Integrated Approach*. J. C. Wagoner (Ed.), SEPM Special Publ (1988).
9. Kennett,J.P. and Srinivana,M.S. Neogene Planktonic Foraminifera (A Phylogenetic Atlas), Hutchinson Ross Publishing Company (1983).
10. Martini,E. Standard Tertiary and Quaternary calcareous nannoplankton zonation. In Proceedings of the Second Planktonic Conference, A.Farinacci (Ed.), pp. 739-785, Tecnoscienza, Roma (1971).
11. NHWOC (Nanhai West Oil Corp.). The Tertiary in the Petroliferous Continental Shelf of the Northern South China Sea, Unpublished reports (in Chinese, 1991).
12. Qing Guoquan. Neogene marine strata in NE Dongsha Islands, discovery and implications. in *Research Reports in Petroleum Geology* II, Nanhai East Oil Corp.(ed.), 640-651 (in Chinese, 1988).
13. Rangin,C., Jolivet,L., Publier,M. and the Tethys Pacific Working Group. A simple model for the tectonic evolution of southeast Asia and Indonesia region for the past 43 m.y. *Bull. Soc. Geol. France*, 21-37(1990).
14. Ru Ke, Zhou Di and Chen Hanzong. Basin evolution and hydrocarbon potential of the northern South China Sea. In *Oceanology of China Seas*, v.2, Zhou Di, Liang Yuanbo and Zeng Chengkui (Eds.), pp.361-372. Kluwer, Dordrecht (1994)
15. Stam,B., Gradstein,F.M., Lioyd,P., and Gillis,D. (1987) Algorithms for porosity and subsidence history, Computers & Geosciences 13, 317-349.
16. Zeng Lin, Huang Lvsheng, and Shu Yu. Zonation of Oligocene Pleistocene calcareous nannofossils for the eastern Zhujiangkou Basin", in *Research Reports in Petroleum Geology* II, Nanhai East Oil Corp.(ed.), 145-181 (in Chinese, 1988).

Proc. 30* Int'l. Geol. Congr., Vol. 11 , pp. 55-59
Wang Naiwen and J. Remane (Eds)
VSP 1997

Application of High-resolution Stratigraphic Correlation Approaches to Fluvial Reservoir

DENG HONGWEN, WANG HONGLIANG
Department of Petroleum, China University of Geosciences, Beijing, 100083, P.R.China
TIMOTH A. CROSS
Department of Geology, Colorado School of Mines, Golden CO 80401 U.S.A.

Abstract

Based on the principle of stratigraphic base level and the techniques of high-resolution chronostratigraphic correlation, this paper focus on recognizing the stratigraphic cycles caused by multi-order base-level cycles and establishing the stratigraphic correlation framework in fluvial deposits. Four intermediate-scale cycles are identified in studing interval in light of detailed-analysis from cores, well logs and seismic data, and the prediction of fluvial reservoirs has been made in each intermediate-scale stratigraphic cycle.

INTRODUCTION

High-resolution stratigraphic correlation approaches based on base-level cycles developed by T.A.Cross are well tested in shallow marine and adjacent coastal plain strata. The purpose of this paper is to identify the sedimentologic and stratigraphic responces to accommodation changes during base-level cycle in fluvial deposits in continental basin and to distinguish multi-order strtigraphic cycles based on those attibutes, establish high-resolution correlation framework and make reservoir prediction.

HIGH-RESOLUTION STRATIGRAPHIC CORRELATION APPROACHES BASED ON THE PRINCIPLE OF BASE LEVEL

Base-level is an abstract potentiometric surface that oscillates above, below and across the physical surface of the earth (Fig.1). As base-level rises above the earth's surface, space is created for sediment accumulation. As base level falls below the earth's surface, erosion occurs.

Stratigraphic cycles observed in stratigraphic record are the products of unidirectional changes of base-level toward maxima and minima of the A/S ratio (the ratio of accommodation and sediment supply). Accommodation change during base-level cycle results in sediment volume partitioning, and the change of the symmetry, thickness, preservation degree , stacking patterns and facies differentiation of stratigraphic

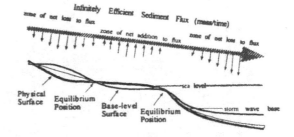

Figure 1. Illustration of base level as a potentiometric surface defining the potential for varying rates of energy and material transfer with respect to the earth's surface in a closed stratigraphic system in which mass, energy and space are conserved

cycles (Cross,1988). Therefore, strtigraphic and sedimentary "traces" of base-level cycle

56

change are preserved in stratigraphic record which can be used to identify stratigraphic cycles.

Identification of a series of stratigraphic cycles is the basis for erecting a high-resolution sequence stratigraphy.Numerous sedimentologic and stratigraphic attributes are used to identify cycles . These attributes may include (a) Vertical successions of facies,.(b) Vertical changes in the physical attributes of a single facies,(c) Stacking pattern of cycles,(d) Geometrical relations of strata. Using one or more of these four criteria, stratigraphic cycles of multiple spatial and durations may be recognized in strata representing any environment. Because the stratigraphic hemicycles correspond to the divisions of base-level cycles, and because the turnaround points of base-level cycles are synchronous at the temporal resolution of the base-level cycles, the correlation are of time surfaces and the stratigraphic cycles are time-strtigraphic units.

APPLICATION TO FLUVIAL RESERVOIR

Identify of multiple-order base-level cycles

The studing area is located in the southeastern part of Chengbei buried hill between Jiyang Depression and Bozhong depression in Bohai bay basin. The upper Tertiary Guantao Formation overlaps the buried hill. It is dominated by fluvial and floodplain facies with a large area and can be divided into two long-term base-level cycles. The upper long-term

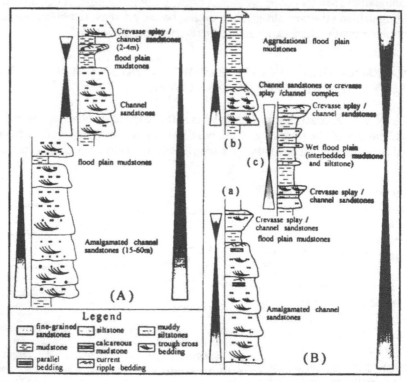

Figure 2. Facies succession motifs in intermediate-scale base-level cycles. The facies succession in small-scale cycles changes as a function of A/S ratio which changes dynamically during intermediate-term base-level cycles.

base-level cycle roughly corresponds to the studying horizon. The basal boundary of this base-level cycle can be identified based on the widespread channel incision and the overlying amalgamated sandstones.

1. Facies succession motifs and stratigraphic cycles from core analysis

Several facies and facies successions can be recognized from cores: braided channel facies succession, low-sinuosity channel facies succession, crevasse splay/crevasse channel facies succession, flood plain facies and aggradational plain facies. These facies successions constitute different small-scale stratigraphic cycles.

a. Low-accommolation cycles: The small-scale cycles deposited in the low-accommodation occur in the lowest part of the studying interval. They are usually in the base of asymmetrical intermediate-scale cycles (Fig.2A). The base-level fall hemicycles are represented by erosional surface caused by the incision of channels,and the base-level rise hemicycles are characterized by amalgamated channel sandstones overlying the erosion surface.

b. Intermediate-accommodation cycle: Small-scale cycles formed in intermediate-accommodation occur in the lower part. They are mainly in the base of symmetrical intermediate-scale cycle. Both base-level rise hemicycle and base-level fall hemicycle are preserved, indicating accommodation space is increasing. But base-level rise hemicycle is much thicker than base-level fall hemicycle. Base-level rise hemicycle comprise superimposed channel sandstones. Base-level fall hemicycle consists of crevasse splay/crevasse splay sitstones (Fig.2 B (a)). Another kind of short-scale cycles formed in intermediate-accommodation occur in the base-level fall hemicycle of symmetrical intermediate-scale cycle. The base-level rise hemicycle is made of thiner fining-up channel sandstones and the base-level fall hemicycle consists of variegated mudstones with thin paleo-soil (Fig.2 B (b)).

c. High-Accommodation Cycles: Small-scale cycles in high accommodation occur in the middle part of symetrical intermediate-cycle, near the rise-to-fall turnaround position (Fig.2 B (c)). They are symetrical, and tend to be

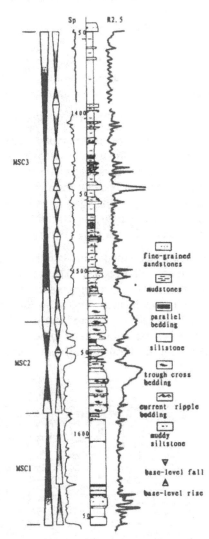

Figure 3. The three intermediate-scale cycles indentified based on the calibration of well log signatures to facies succession described in cores. Stacking patterns of stratigraphic units of intermediate-scale cycles change as a function of A/S ratio which changes dynamically during long-term base level cycles.

58

dominated by floodplain grey-green mudstones and calcareous mudstones which indicate the maximum accommodation.

2. Stratigraphic cycles from well logs

Intermediate-scale cycles can be recognized effectively based on the stacking patterns of short-scale cycles(Fig.3). The base-level rise hemicycle of the upper long-term base-level cycle consists of MSC1,MSC2 and the base-level rise hemicycle of MSC3. From the basal erosion surface of MSC1 to the turnaround point of MSC3, stacking pattern of short-scale cycles, the change of cycle symmetry, increase of floodplain mudstones, and the preserve of crevasse splays are evidence of increasing accommodation. Base-level fall hemicycle of MSC3 and MSC4 comprise the base-level fall hemicycle of the long-term base-level cycle. The increase of variegated mudstones deposited by dominantly vertical accretion, and the occurrence of soil indicate decreasing accommodation.

3. Stratigraphic analysis of seismic data

Seismic sequence stratigraphy analysis is abtained by converting stratigraphic interpretation from well logs and cores into seismic profiles and also the following criteria can be used to identify base-level cycles in seismic profiles of Guantao Formation:(a) Channel incision resulted from by reginal base-level fall and the discrete onlapping in the base of the upper Guantao Formation generated by the begining rise of regional base-level. (b) Better-continuous, high-amplitude, lower-frequence reflection correlated to rise-to-fall turnaround of longer base-level cycles. (c) Regionally significant change in seismic facies.

In low accommodation conditoin, multiple-storied, amalgamated braided channel sandstones with bigger thickness and the blanked-like distribution character in a large area, tend to generate good impedance contrasts at the base and top with the underlying and overlying mudstones. The more homogeneous and less mud-bearing sandstones usually produce weak or a transparent reflection zone. In high accommodation condition, channelbelt sandstones tend to be less multiple-storied or isolated and ribbon-like in plan view with the limited areal extention. The seismic reflections produced by this kind of

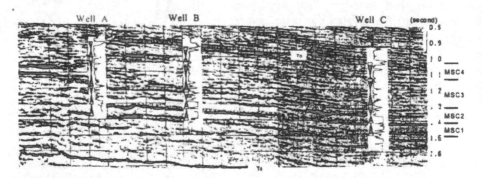

Figure 4. Seismic intervals indentified based on seismogram synthesis and seismic facies analysis, which correspond to the intermediate-scale cycles determined from cores and well logs. MSC1: weak-transparent reflections, disperse onlapping terminations of the basal boundary of the cycle. MSC2: upper part: weak-reflections with discontinuous, midium-low amplitude reflections; middle part: continuous, high-amplitude reflections; lower part: same as MSC1. MSC3 upper part: weak-reflections and weaklly with low-angle, medium-amplitude and high frequence reflections; lower part: weak-reflections with alternating horizontally high-medium continuous, high-amplitude and lower-frequence reflections.

channelbelt should be non-continuous, high-mediam amplitude, lower-frequence reflections in weak or nearly transparent reflection background generated by thicker floodpain mudstones. The deposits of cravasse splay/ cravasse channel complexes produce lower-angle, sometimes shingled mediam-high amplitude reflections in weak reflection background produced by floodplain mudstones.

Seismic stratigraphic cycles which correspond to the intermediate-term cycles in well logs can be identified. Each cycle can correspond to distict seismic facies (Fig.4)

Fluvial facies reservoir prediction

Guantao Formation underwent the evolution from braided channels to low-sinuosity channels. In the lower part reservoirs are braided channel sandstones of MSC1 and MSC2 with high homogeneity and broad latural extend. The dominated braided channel of MSC1 and MSC2 are substituted by low-sinuosity meandering channel-floodplain environment of MSC3 and MSC4 with the base-level rise and accommodation space increases. The channel sandstones are isolated, with poor connectivity and heterogeneity.

The reservoir of meandering channelbelt sandstones in MSC43 and MSC4 are regional regularly distributed(Fig.5). Generally, in the same geographic position, the position where multiple-channels occur during base-level rise time are substituted by floodplain and crevasse splays during base-level fall time, and *vice versa*. The effect of topographic relief and the differential compaction may contribute to the characters of reservoir distribution.

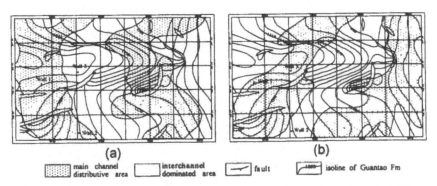

Figure 5. Map showing the main channel distribution in MSC3 (a. base-level fall hemicycle, b.base-level rise hemicycle) ,delineated based on 3D seismic data. The location of main channels geographically changes during the base-level cycle.

CONCLUSIONS

This study contributes to the understanding of the responses of fluvial deposits to base-level cycles and indicates the high-resolution correlation approaches can be also used in continental strata.

REFERENCES

1. Cross T. A. Controls on coal distribution in transgressive-regressive cycles, Upper Cretaceous, Western Interior, U.S.A.In:Wilgaus C K, et al. *Sea-level changes: An integrated approach*. SEPM Special Publication, 42, 1988. 371-380

Proc 30ᵗʰ Int'l. Geol. Congr., Vol. 11 , pp. 61-66
Wang Naiwen and J Remane (Eds)
© VSP 1997

Carboniferous Sequence Stratigraphy and Oil and Gas in Tarim Basin, Northwest China

JIAYU GU

Research Institute of Petroleum Exploration and Development, CNPC, CHINA

Abstract

Tarim Basin is a large Petroliferous basin with rich petroleum resources and good exploration prospection. Carboniferous is one of the main oil-bearing strata in the basin. On the basis of strata correlation and analysis of sequence stratigraphy, Carboniferon in Tarim Basin can be divided into 8 lithologic members whch is grouped into 3 sequsnces. According to analysis of sequence, we consider that source rock is about 400m in the southwest area. Reservoir occurs in the lower part of transgressive system tract with 200m thick (named Donghe sandstone) in delta sandbody of high stand system tract. Caprock in Carboniferous is well develped and is one of the important regional cover. Oil and gas derived from low paleozoic source rock accumulate under the cover to form middle and large oil and gas fields.

Keywords: Carboniferous, sequence boundary, lithologic member, Lowstand system tract, highstand system tract, transgressive system tract, sea-level

Tarim basin is a large Petroliferous basin with rich petroleum resources and good exploration prospection. Reserve for oil and gas is up to 19. 1 billion tons, which account for 1/5 of the total reserve in China. Carboniferous is one of the main oil-bearing strata in the basin and its reserve accounts for 1/4 of that in the basin. Tazhong-4 oil field and Donghetang oil field etc. have been found in the strata. By studying on Carboniferous sequence stratigraphy we shall recognize the distribution characteristics of source rocks, reservoir, and cap rocks and to find more large oil and gas field effectively and precisely.

TECTONIC SETTINGS FOR CARBONIFEROUS SEDIMENTATION

South Tianshang marine basin, which is fomed by the extension of the northern margin of Tarim basin in Silurian period, begam shrinking in carboniferous period. The rock association shows that continental sediment predominates in some area. Widespreading gypsum-bearing limestone or gypseous mudstone indicates that nearly enclosed marine basin is formed in south Tanshang in carboniferous period. In South margin of the basin the Middle Kunlun mountain and Tarim continent post Silurian is integrated because of the collision. The southwest margin become a passive continental margin. Correlation of cross sections show that carboniferous is a transgressive series.

The basin consists of southwest Tarim pericratonic depression and Tarim cratonic depression. Southwest Tarim pericratonic depression with 121000km² area, which is located in the southwest of Tarim basin, mainly deposits the sediments of open-platform facies, platform margin facies and restricted deep slope facies, the thickness ranges from 1000 to 2000m. Tarim cratonic depression is a stable subsided district with about 400000km² area, deposits the sediments of littoral and shallow-water platform facies, the thickness ranges from 500 to 1200m. What should be emphasized is that lower carboniferous consists of greyish white littoral facies sandstone with 100 to 200m thick, and of epicontinental sea facies carbonate and clastic rock(Fig. 1).

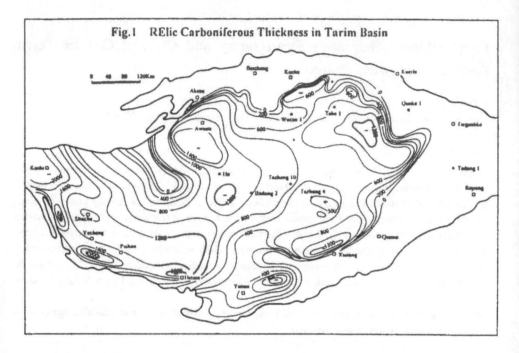

Fig.1   RElic Carboniferous Thickness in Tarim Basin

CARBONIFEROUS SEQUENCE STRATATIGRAPHY OF TARIM BASIN

Stata distribution and division of lithologic members

Carboniferous strata mainly distributes in the center and U-shaped area in west of the basin, the thickness increases from east to west. The thickness in the east and low-uplift area from well Manxi to Tazhong are 100m to 400m and 400m respectively. The thickness in Manjar depression about 400-600m, well He 2 and the front of Kunlun mountain in the southwest of the basin is 800-1400m.

On the basis of strata correlation and sequence stratigraphy analysis, carboniferous in Tarim basin can be divided into 8 lithologic members. which is grouped into 3 sequences. The 8 lithologic members and 3 sequences can be correlated to the basinal extent.

1st lithologic member (lime mudstone), Greyish white, grey, yellowish brown limestone, mudstone and sandstone change rhythmically in the lithologic member. The lithologic member in Tabei area is mainly composed of micrite and partly bioclastic limestone with 12-120m thick.

2nd lithologic member (sandy mudstone), The main characteristics is that greyish white, yellowish sandstone, siltstone interbeds with grey, greenish grey, dark purplish brown pelitic sandstone.

3rd lithologic member (upper mudstone), Light grey, light brown, dark purple mudstone interbed with thin siltstone, lime mudstone and pelitic siltstone. The thickness of the member is 28. 0 to 50. 0m.

4th lithologic member (bimodal limestone), Rock types include grey micritic limestone, marl, which interbed with thin mudstone. This lithologic member are widespread with stable thickness ranging from 9. 5 to 20. 0m. The member may be correlated in the whole basin.

5th lithologic member (lower mudstone), Light grey, dark purple and brown mudstone, pelitic siltstone interbed with thin siltstone. Thick gypsolith distributes in Manjar depression. Gypsum layer and pelitic gypsolith are well developed in well He 2. The thickness is 43. 0 to 110. 0m.

6th lithologic member (lime dolomite), Rock types include micritic bioclastic dolomitic limestone, micriti dolomite, micritic alga clastic dolomitic limestone etc. , and secondary calcarenite, and bioclastic sparite. Dolomite and limestone change rhythmically with 25-40m.

7th lithologic member (mudstone), Light grey mudstone and secondary siltstone are the main types of rocks in the member. The member is strictly controlled by sedimentary facies and lithology change greatly. The thickness is 20. 0 to 45. 0m.

8th lithologic member (Donghe sandstone), The lithologic member is firstly found in Donghetang area and named "Donghe sandstone". Rock types include brownish grey, light grey fine sandstone, medium sandstone and silty fine sandstone with good sorting.

Determination of sequence boundary and division of sequences

Van Wagoner et al. (1990) proposed that a depositional sequence is bounded by unconformities and their correlative comformity. The key to divide sequences is to determine sequence boundies. The strata unit between two sequence boundaries is a sequence when sequence boundaries are determined.

Method to determine sequence boundarys

1)Outcrops show that lower carboniferous unconformably overlays Devonian and Silurian etc. strata of different geologic time.

2)The Lithology, cementation degree and mineral component of sandstone are different, though Carboniferous is conformable with Devonian in some area.

3)In Xiaohaizi outcrop section, a layer of kaolin derived from feldspar weathering. is found in the unconformity surface of carboniferous bottom. It widely distributes and the strike length on the outcrop is 10m or so. The layer is recognized as paleosol layer.

4)Carboniferous cut the lower strata by an angle on seismic section. Carboniferous onlaps the paleouplift and its top is truncated in Tabei and Tadong uplift; Lower Carboniferous obviously eroses and fills the underlying strata.

5)Abrup change of lithology and lithologic facies, break of paleotologic species, and the change of sedimentary cycles are used to determine sequence boundary on log sections. Well log shows that carboniferous overlays underlying strata of different geologic time such as Devonian, Silurian, Ordovician directly in Tabei uplift and as Ordovician in Tazhong. Fluvial sediment directly overlays the mudstone of marine platform facies at the margin of uplift area.

6)Various log curves have abrupt changes. The curve value change from low to high abruptly and acoustic log curves show that there is obvious velocity change.

Division of sequence

According to the types of sequence boundaries, Carboniferous bottom is type- I sequence boundary and there is two type- II boundaries inside Carboniferous. Carboniferous may be divided into three sequences. The sequence types show that when the lower sequence is formed, the rate of global sea level fall exceeds the deposition rate at depositional coastal break; sea water retrograde to the position below depositional coastal break; streams rejuvenates in widely-spreading shelf; incised valley is formed in continental slope and there is deposition of basin-floor fan. So sequence- I in the basin may be divided into three system tracts; lowstand system tract, transgressive system tract, highstand system tract.

Sea level fall rate is a little lower or equal to the deposition rate in depositional coastal break when middle Carboniferous (sequence II) and upper Carboniferous (sequence III) are formed. It shows that there is no sea level fall at depositional coastal break. Sea level fall rate at the center of the basin is relatively little. So what we see in the basin is the deposition of transgressive system and highstand system rather than lowstand system tract.

CHARACTERISTICS OF SYSTEM TRACTS

It is difficult to divide sequences in basin main body because the structure of Tarim basin at Paleozoic is relatively stable and seismic reflection surface is relatively straight. Restriction inversion of seismic section is carried out on the basis of well logging, coring data in order to divide sequences and recognize different sedimentary system tracts directly.

Firstly seuqnces are divided on logging sections to determine relation between well depth and time of seismic section. Strata determination data are listed out to construct the relation between acoustic layer speed and lithologic characteristics. Finally acoustic parallax, acoustic layer speed, seismic record and lithologic to draw out the sequences, trace and divide them on seismic sections. Inside sequence, on the basis of the study of section reflection sturcture, system tracts are distinguished and seismic facies are divided. According to the relation of seismic and sedimentary facies, horizonal distribution of sedmitary facies is determined.

Division of system tracts

As proposed by Posameutier, H. W et al. (1988), sedimentary system tract is the product of relative sea level change, though sedimentary system tract only indicates the position of strata on sections, In lowstand system tract, basin-floor fan, slope fan lowstand wedge are formed then sea level rapidly falls from A to B, B slowly falls to C and C slowly rise to D.

Transgressive system tract is formed when sea level rapidly rises from D to E.

Highstand system tract is formed when sea level slowly rises to F and slowly falls to G.

The above shows that the key to distinguish the three system tracts is to recognize the initial point (surface) of sea level rapidly falling, initial rapidly transgressing point (surface) and the largest sea flooding surface.

Sea level change is reflected in two aspects in geological history: marine sediment transgressing to ward and retrograding away from continent and the lithologic change. We find the first onlap point and take it as initial rapidly transgressing point (surface) and then trace it to the farthest point of transgressive strata to determine the largest sea flooding surface. The sedimentary unit between the first onlap point and the sequence bottome is lowstand system tract and the sedimentary unit between the first onlap point and the farthest point of transgressing is transgressive system tract. The sedimentary input volume exceeds the accommodation provided by slowly transgressing to form regradation and the set of sediment is highstand system tract.

Characteristics of different system tracts

Sedimentary characteristics and associations of identical system tracts of the three sequence in Tarim basin have great difference.

Lowstand system trace of the three sequence is only distributed on Wushi section and it does not exist inside the basin. The characteristics of lowstand system tract of sequence I is that sediment of basin bottom fan and dark marginal mudstone, which is mainly composed of pebbled sandstone and sandstone, deposit on the erosion surface. Sediment inside the basin is thin and the main sedimentary types are fluvial lag deposit and fill deposit. . Shelf slope fan, which is composed of sandstone interbeded by shale or sandy conglomerate, is improtant in the lowstand system tract of sequence I and Ⅱ, where slump, corrugition, involution sturctures which are resulted from gravity are well developed(Fig. 2)

Transgressive system tract has widespread occurrence in Tarim Basin and the sediment is thick. Sediment change regularly from west to east due to transgressing from west to east.

At the initial transgressive period, shelf carbonate predominates at the west margin of the basin due to sea water influx from west to east. Sandstone and siltstone of coastal shelf predominate in the basin due to the shallow water, topography relief, and enough terrestrial clastic supply. With transgress becoming more widely, sediment fines from sandstone to siltstone and pelitic sandstone gradually. Car-

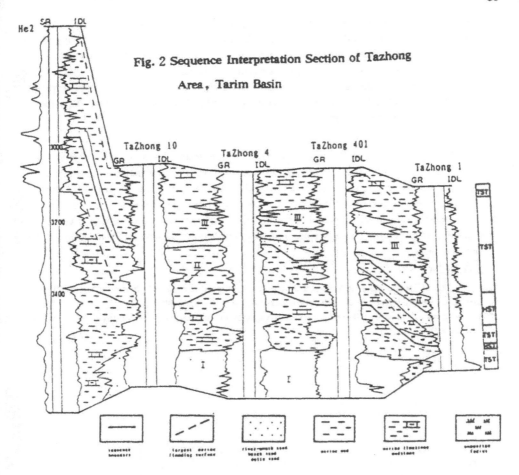

Fig. 2 Sequence Interpretation Section of Tazhong Area, Tarim Basin

bonates, such as micrite calcarenite and partially dolomite, deposit on the extensive platform because the terrestrial matter retrogrades toward continent and poor terrestrial supply after transgress achieves its highest level. In transgressive system tract of sequence Ⅲ, due to slowly transgressing, influence of Hercynian movement to the basin and warm weather, deep-water mudstone deposits at the west margin at initial and middle transgressive period, while greenish grey mudstone and gypseuous mudstone, gypsolith interbeded by dolomite deposit inside the basin. Sand-rich delta and coastal sediment deposit in Tadong Area due to enough terrestrial clastic supply resulting from uplifting and high topographic relief.

Sediment of carboniferous highstand system tract in Tarim basin is relatively thin. The strata may be eroded afterward. Shelf silty mudstone is predominant in the west and the main sedimentary rocks include purplish brown pelitic siltstone, silty mudstone of tidal flat and partially low-energy intertidal and subtidal zone deposits.

PREDICTION OF SOURCE, RESERVOIR AND COVER

Deposition of shallow-water epicontinental sea named platform in sedimentology, predominates in Tarim basin in Carboniferous. It makes source rock poor developed on the whole except in the southwestern area where the source rock is 400m or so thick in 3 sequences.

Reservoir occurs in the lower part of transgressive system tract which is mainly composed of coastal clastic sandstone and delta sandstone body tends to 200m thick. The reservoir has good physical feature. Highstand system tract is not the major reservoir due to erosion or bad physical feature.

Caprock is carboniferous is well developed and is one of the important regionl cover in the whole basin. It is with widely distribution great thick and good seal ability. The types of rock include gypsolith, mudstone with 40-360m thick and which acount for 16—86% of Carboniferous strata. They form in the middle to late period of transgressive system.

Conclusions

Carboniferous in Tarim basin Can be divided into three sequences, 23 parasequences, parasequence stacking patterns is also divided into progradation, retrogradation and aggradation. Lowstadn system tract is not found in the sequences except sequencE Ⅰ. The beach sedimentary bodies which forms in the initial period of transgressive system tract are good reservoirs. The middle and upper part of transgressive system tract which is composed of grey mudstone and micrite are good regional cap-rock. Oil and gas derived from lower Paleozoic source rocks accumulate under the cover to form middle and large oil and gas fields.

REFERENCES

1. T. S. Loutit, J. Hardenbol, P. R. Vail, and G. R. Boum, Condensed sections; The key to age and correlation of continental margin sequences, *SEPM Special pblication* 42, P. 183—217(1988).
2. Van Wagoner, J. C. et al, Siliciclastic sequence stratigraphy in well logs, cores and outcrops; *AAPG Methods in Exploration Series* 7(1990).

Proc 30ᵗʰ Int'l. Geol. Congr., Vol. 11 , pp. 67-73
Wang Naiwen and J. Remane (Eds)
© VSP 1997

Neoproterozoic Acritarch Biostratigraphy of China

YIN LEIMING

Nanjing Institute of Geology and Paleontology, Academia Sinica, Nanjing, 210008

YIN CHONGYU

Institute of Geology, Chinese Academy of Geological Sciences, Beijing, 100037

Abstract

According to studies of the acritarch biostratigraphy in China, the evolution of Neoproterozoic acritarchs roughly involves four stages, from older to younger, including the pre-Doushantuo Stage, the Doushantuo Stage, the Dengying Stage and the Meishucun Stage. Most of the species with large size and complex ornament structures occured in the first stage, flourished in Doushantuo stage and disappeared near the time of the main Ediacarian animal radiation. Following the early Cambrian transgression, the succedent radiation of acanthomorphic acritarchs dominated by the genus *Michystridium* marks the beginning of Cambrian.

Keywords: Neoproterozoic, acritarch biostratigraphy, acanthomorphic acritarch, evolution

INTRODUCTION

Neoproterozoic (ca.1000-570Ma) strata are widely exposed in China. Its succession is separated into three major divisions by the glaciogenic Nantuo Formation. The pre-Nantuo early Neoproterozoic (ca.1000-720Ma) succession is characterized with carbonate and clastic rock in North China, northern South China. The volcanogenic sediments are mainly distributed in Northwest China and southern South China. The glaciogenic Nantuo Formation, which is most likely equivalent to the Varanger tillite in Europe and the Marinoan ('upper') tillite in Australia but radiometrically dated as old as ca. 720-680Ma, are widely distributed over the Yangtze Platform in South China. The post-Nantuo late Neoproterozoic(ca.680-570Ma) succession is represented by stratigraphic sequence as named as the Doushantuo Formation and the Dengying Formation in South China.

The Doushantuo Formation is 100-300m thick, thickest up to 800m, consisted of sandstone and shale, but in depression area more carbonate and silicalite appear in its upper part. Phosphate sediments of the Doushantuo Formation are exposed in southern Shaanxi Province, western Hubei Province, western Hunan Province and northern Guizhou Province.

The Dengying Formation is commonly composed of dolomite and dolomitic limestone, 200-900m thick. Siliceous rock and chert lenses are often seen in its upper part. The Dengying Formation is overlain by the lowest Cambrian yielding many small-shelly fossils.

Since the late sixties, more and more palaeontological evidences have been found from Neoproterozoic strata in China. Of which, many fossil records are well known in China and abroad. We would like, in the short paper, to make a summarization of Neoproterozoic acritarch biostratigraphy in China.

NEOPROTEROZOIC ACRITARCH BIOSTRATIGRAPHY IN CHINA

The early evolution of acritarchs in China roughly involves four stages. From older to younger, they are as follows: the pre-Doushantuo Stage, the Doushantuo Stage, the Dengying Stage and the Meishucun Stage.

1. The pre-Doushantuo Stage

The acanthomorphic acritarchs, including *Shuiyousphaeridium*, *Tappania*, *Trachyhystrichosphaera* and *Cymatiosphaeroides* have been found from the Ruyang Group in the Shuiyou section in Yongji County, Shanxi Province and Maijiagou section in Mianchi County, Henan Province (Guan et al., 1988; Yan and Zhu, 1992; L.Yin in press; Yin, C. 1995; in press). In addition, the spheromorphic acritarchs of the Ruyang Group also display distinctly morphological features, such as occurrence of obvious round opening, striated ornamentation on the vesicle wall, e.g. *Tasmanites*, *Dictyosphaera*, *Simia* etc. Based on size, morphology, surface texture and characters of processes, most of species found in the Ruyang Group are more or less similar to those found in Neoproterozoic. However, the age of the Beidajian Formation of the Ruyang Group is determined by K/Ar method of glauconite to be 1183+73Ma. In addition, according to the recent work on stable carbon isotopes by Xiao and others (in press), a near zero to slightly negative C secular trend in carbonates of the overlying Luoyu Group suggest a Mesoproterozoic (ca. 1600-1000Ma) to earliest Neoproterozoic age.

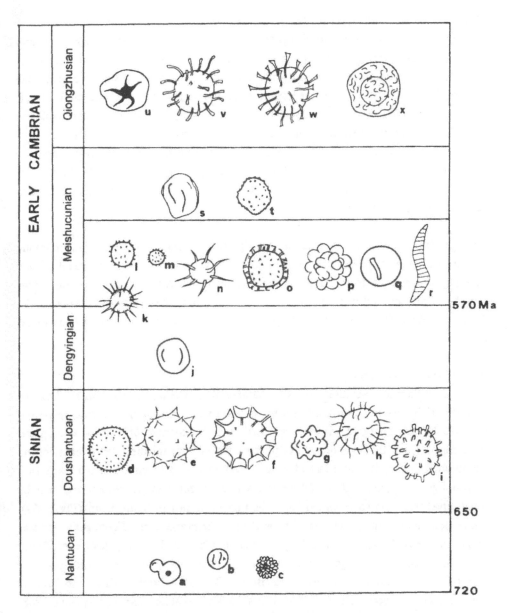

Figure 1. Selected acritarchs from the Sinian-Cambrian transition on the Yangtze Platform a. *Leiomimuscula* sp.; b. *Protoleiosphaeridium* sp.; c. *Sphaerocongregus variabilis*; d. *Echinosphaeri-dium maximum*; e. *Asterocapsoides sinensis*; f. *Tianzhushania spinosa*; g. *Eotylotopalla delicata*; h. *Ericiasphaera spjeldnaeii*; i. *Papillomembrana compta*; j. *Leiosphaeridia* sp.; k. *Micrhystridium regulare*; l. *Filisphaeridium echinulatum*; m. *Filisphaeridium minutum*; n. *Micrhystridium amplia-tum*; o. *Paracymatiosphaera irregularis*; p. *Symphysosphaera* sp.; q. *Eoaperturilites monoscissus*; r. *Megathrix longus*; s. *Leiosphaeridia asperata*; t. *Lophosphaeridium* sp.; u. *Archeodiscina umbonu-lata*; v. *Skiagia orpicularis*; w. *Skiagia ornata*; x. *Annulum squamaceum*.

The indisputable acritarch assemblage of the early Neoproterozoic (ca. 1000-850Ma) is represented by those from the Qingbaikou series in the Yanshan Range in central North China Platform. Qingbaikou acritarch assemblage is dominated by spheromorphic acritarchs. Most of them are not ornamented single vesicle or cell - like colony. In about 850-700Ma, the aspect of acritarch assemblages was more or less changed. Several new forms, such as *Sinianella, Pololeptus, Trachyhystrichosphaera* and *Asterocapsoides*-like fragments occur in the acritarch assemblages from the Liulaobei and Shijia Formations in Anhui Province. Futurmore some intercontinentally distributed forms of this geological epoch, e.g. *Trachyhystrichosphaera, Germinosphaera*, etc., have been obtained from the Dongjia Formation of Lushan region in Henan Province.

The glaciogenic Nantuo Formation was probably related to the frigid climatic conditions, phytoplanktonic organisms decreased evidently in both quantity and diversity. Even those from the interglacial shale of the Datangpo Member in eastern Guizhou Province and western Hunan Province, such as acritarchs of *Leiominuscula, Protoleiosphaeridium, Sphaerocongregus* etc., are generally very small in size(Fig.1, a-c).

2. The Doushantuo Stage

The end of Nantuo glaciation causes the temperature risening up to a warm clime. In addition, rich nutrients brought from the deep sea water by a major oceanic upwelling (see Cook and Shergold, 1984) and carried by meltwater of a new widespread trangressive favor the development of phytoplanktonic organisms. The diversity of the acanthomorphic acritarchs reached a new level in the Doushantuo epoch. Based on morphological features, about 20 genera and 30 species of acanthomorphs can be determined from the Doushantuo cherts and phosphorites in South China(Fig.1, d-i). Some dominant forms, such as *Tianzhushania, Ericiasphaera, Echinosphaeridium, Megahystrichosphaeridium, Trachyhystrichosphaera* and *Briareus*, have been found from about equivalent age succession in the other parts of the world, e.g. the Pertatataka Formation of the Amadeus Basin, Central Australia (Zang and Walter, 1989, 1992); the Scotia Group of Prins Karls Forland, Svalbard (Knoll and Ohta, 1988, Knoll, 1992); the past Riphean - early Vendian strata in Siberian Platform (Sokolov and Iwanovsky, 1990); the Biskopas Conglomerate in the Hedmark Group, southern Norway (Vidal, 1990); the Vendian Chichkan Formation in southern Kazakhastan (Sergeev,1989) and the Infrakrol Formation of the Lesser Himalaya, India (Tiwari and Knoll, 1994). As known data, it is undoubted that the acanthomorphic acritarch assemblage from the Doushantuo Formation of the Yangtze Platform is most remarkable to displaying the aspect of phytoplankton evoluation during the geological period just before the Ediacaran faunal radiation.

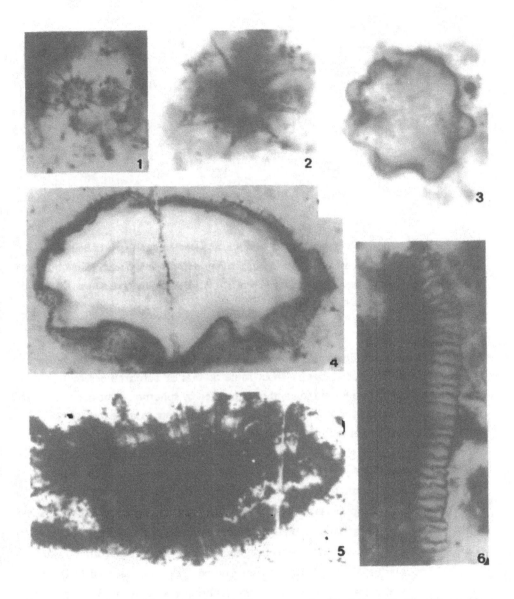

Figure 2. Acanthomorphic acritarchs and filamentous microfossil from the Doushantuo Stage and the Meishucun Stage in South China. All of them were observed in petrographic thin sections. 1. *Micrhystridium regulare* L. Yin, Kuanchuanpu Formation, Ningqiang region, Shaanxi. x 1000; 2. *Micrhystridium ampliatum* Wang, Yanjiahe Formation, Miaohe village, Ziqui, Hubei, x 1000; 3. *Eotylotopalla dactylosa* L. Yin, Doushantuo Formation, Liuxi section, Gaojiayan, Changyang County, Hubei, x 390; 4. *Echinosphaeridium maximum* (L. Yin) Knoll, Wangfenggang section, Liantuo, Yichang County, Hubei, x 200; 5. *Tianzhushania polysiphonia* C. Yin, Wangfenggang section, Liantuo, Yichang County, Hubei, x 140; 6. *Megathrix longus* L. Yin, Yurtus Formation, Aksu region, Xinjiang, x 120.

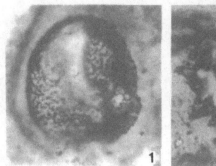
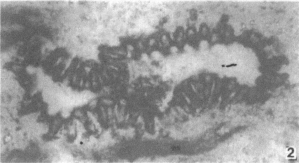

Figure 3. Morphologically complex acritarchs from the Doushantuo Stage in South China. 1. *Lophosphaeridium* sp., Doushantuo Formation, Tianjiayuanzi section, Yichang County, Hubei, x 410; 2. *Papillomembrana compta* (Spjeldnaeii) Vidal, Doushantuo Formation, Xiaofenghe section, Yichang County, Hubei, x 210.

3. The Dengying Stage

From the Doushantuo Formation upwards to the overlying Dengying Formation, the aspect of acritarch assemblage is about same as that of acritarch assemblage from shale of the Doushantuo Formation and is dominated by spheromorphic acritarchs. But, most species of large, morphologically complex acritarchs disappeared at about the time of Ediacaran faunal radiation(Fig.1, j).

4. The Meishucun Stage

The Lowest Cambrian microflora in the Yangtze Platform is dominated by occurrence of *Micrhystridium* Complex, as well as *Paracymatiosphaera, Megathrix*, etc (Fig.1, k-r; Fig.2, 1-2,6). The horizon is correspondence with the small shelly *Anabarites-Circotheca-Protohertizina* Zone and is roughly coeval with the *Asteridium tornatum-Comasphaeridium velvetum* biozone in East Europe. They are assumed to indicate a stratigraphic level close to the Precambrian-Cambrian boundary marked by the first appearance of *Phycodes pedum* in eastern Newfoundland.

As mentioned above, the evolutionary stages of the early acanthomorphic acritarchs offer a criterion of endemic and worldwide correlation of the Neoproterozoic strata.

REFERENCES

1. P. J. Cook and J. H. Shergold. Phosphorus, Phosphorites and Skeletal evolution at the Precambrian-Cambrian Boundary. *Nature*, 308, 231-236 (1984).

2. Guan Baode, Geng Wuchen, Rong Zhiquan and Du huiying (Eds). *The middle and upper Proterozoic in the northern slope of the Eastern Qinling ranges, Henan, China*. Henan Science and Technology Press, Zhengzhou.. (1988, in Chinese, with English abstract).

3. A. H. Knoll. Vendian microfossils in metasedimentary cherts of the Scotia Group, Prins Karls Forland, Svalbard. *Palaeontology*, 35:4, 751-774 (1992).

4. A. H. Knoll and Y. Ohta. Microfossils in metasediments from Prins Karls Forland, Western Svalbard. *Polar Research*, 6, 59-69 (1988).

5. V. N. Sergeev. Microfossils from transitional Precambrian-Phanerozoic strata of central Asia. *Him. Geol.*, 13, 269-278 (1989).

6. B. S. Sokolov and A. B. Iwanovsky (Eds). *The Vendian System Vol. 1 Paleontology*. Springer-Verlag, 154-195 (1990).

7. Meera Tiwari and A. H. Knoll. Large Acanthomorphic Acritarchs from the Infrakrol Formation of the Lesser Himalaya and their Stratigraphic Significace. *Palaeontology*, 37:2, 204-219 (1994).

8. G. Vidal. Giant acanthomorph acritarchs from the Upper Proterozoic in southern Norway. *Palaeontology*, 33, part2, 287-298 (1990).

9. Yan Yuzhong and Zhu Shixing. Discovery of acanthomorphic acritarchs from the Baicaoping Formation in Yongji, Shanxi and its geological significance. *Acta Micropalaeontologica Sinica*. 9:3, 267-282 (1992, in Chinese, with English abstract).

10. Yin Chongyu and Gao Linzhi. The early evolution of the acanthomorphic acritarchs and its biostratigraphical implication in China. *Acta Geologica Sinica*. 69:4, 360-371 (1995, in Chinese); 9:2, 193-206 (1996, in English).

11. W. L. Zang and M. R. Walter. Latest Proterozoic plankton from the Amadeus Basin in central Australia. *Nature*, 337, 642-645 (1989).

12. W. L. Zang and M. R. Walter. Late Proterozoic and Cambrian microfossils and biostratigraphy, Amadeus Basin, central Australia. *Mem. Ass. Australas. Palaeontols* 12, 1-132 (1992).

Proc. 30ᵗʰ Int'l. Geol. Congr., Vol. 11 , pp. 75-84
Wang Naiwen and J. Remane (Eds)
© VSP 1997

Tremadoc Trilobites from the Mungog Formation, Yeongweol, Korea

DONG HEE KIM AND DUCK K. CHOI

Department of Geological Sciences, Seoul National University, Seoul 151-742, Korea

Abstract

The Mungog Formation consists predominantly of carbonate with lesser amounts of shale, representing a shallow marine environment. It is divided into four members based on the association of dominant lithofacies such as ribbon rock, grainstone to packstone, flat-pebble conglomerate, and marlstone to shale facies. The basal member, ca. 45 m thick, consists mainly of ribbon rock and grainstone to packstone with intercalations of thin flat-pebble conglomerate beds. Trilobites occur at the lowermost several-m-thick interval of the basal member. The lower member is recognized by the occurrence of a thick (30-35 m in thickness) sequence of dolostone and no trilobites have been recovered from this member yet. The middle member, 35-60 m thick, is characterized by alternations of ribbon rock and flat-pebble conglomerate lithofacies with occasional intercalations of grainstone to packstone beds. *Kainella*, the only known trilobite from the middle member, occurs at the lowest bed (ca. 30 cm thick) of the member. The upper member, 50-60 m thick, comprises ribbon rock, grainstone to packstone, flat-pebble conglomerate, and marlstone to shale facies. This member yields relatively abundant and diverse trilobites along with brachiopods, crinoids, ostracods, and fossils of uncertain zoological affinity. Fossil occurrences in the Mungog Formation are confined to the three stratigraphically separated intervals, which are referred to the lower, middle, and upper fauna, respectively. The lower fauna consists dominantly of *Yosimuraspis* and subordinately of *Jujuyaspis* and *Pseudokainella* and indicates an early Tremadoc in age. Comparable faunas are well represented in North China. The middle fauna represented solely by *Kainella* can be correlated with the middle Tremadoc faunas of North America and Argentina. The upper fauna comprises mainly cosmopolitan trilobite taxa including *Micragnostus, Asaphellus, Shumardia, Hystricurus, Apatokephalus*, and *Dikelokephalina*, with some endemic species such as *Hukasawaia cylindrica* and *Koraipsis spinus*. It is closely comparable to the Dumugol fauna of the Duwibong sequence in Korea and late Tremadoc faunas of North China and Australia.

Keywords: trilobites, Ordovician, Tremadoc, correlation, Mungog Formation, Korea

INTRODUCTION

The Mungog Formation occupies the Lower Ordovician portion of the Cambro-Ordovician Joseon Supergroup in Yeongweol, Korea. The formation has been known to yield diverse invertebrate fossils. Yosimura [30] was the first to report the occurrence of fossils from the Mungog Formation and subsequently Kobayashi and Kimura [18] recorded the occurrence of Early Ordovician graptolites from the formation. Kobayashi [15] described three trilobite taxa from the Mungog Formation. All of these materials were compiled in a monographic study of the Mungog fauna by Kobayashi [16]. During the past three decades no paleontological studies on the macroinvertebrate fauna of the Mungog Formation had been carried out, until recently additional fossils from the Mungog Formation were documented [11, 12, 22]. These findings significantly modify and supplement the previous stratigraphic scheme of the Mungog Formation. This report summarizes a refined lithostratigraphic and biostratigraphic framework of the Mungog Formation and also discusses the correlation of the Mungog trilobite faunas with coeval ones elsewhere.

GEOLOGIC SETTING

The Cambro-Ordovician sedimentary rocks in South Korea, which are collectively called the Joseon Supergroup, are exposed in the northeastern part of the Ogcheon Fold Belt (Fig. 1). The sediments consist mainly of limestone and dolostone and subordinately of sandstone and shale, which were presumably deposited in a continental margin-type depression fringing the early Paleozoic Sino-Korean block. Kobayashi et al. [19] recognized five types of sequences within the Joseon Supergroup, each with distinct lithologic succession and geographic distribution; i.e., Duwibong, Yeongweol, Jeongseon, Pyeongchang, and Mungyeong sequences. The Yeongweol-sequence Joseon Supergroup was divided into five formations; namely, Sambangsan, Machari, Wagog, Mungog, and Yeongheung Formations in ascending order [30]. The lower three formations were assigned to the Cambrian, while the upper two to the Ordovician [17].

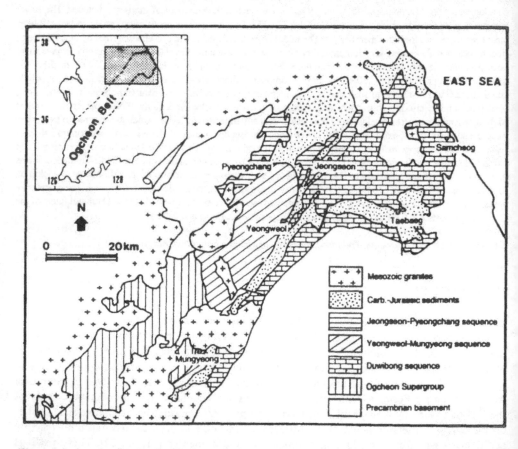

Figure 1. Index map showing the distribution of the lower Paleozoic Joseon Supergroup in the northeastern part of the Ogcheon Belt, Korea. All of the trilobites considered in this study are collected from the Yeongweol sequence located in the area between Yeongweol and Pyeongchang.

The Mungog Formation is composed of diverse lithotypes [6, 21, 22] and attains a thickness of up to 200 m. It rests conformably on the Wagog Formation which is a thick (ca. 250 m in thickness) sequence of light gray to gray massive dolostone. The Mungog Formation grades conformably into the Yeongheung Formation which consists mainly of dark gray dolostone and limestone. Recent sedimentological studies have shown that the Mungog sediments were deposited in a shelf environment ranging from tidal flat [21] to somewhat deeper subtidal environment [6].

LITHOSTRATIGRAPHY

The Mungog Formation was established by Yosimura [30], who conceived the formation to be subdivided into three parts: the lower part comprises bluish gray limestone, marl, and greenish shale; the middle part is composed predominantly of gray massive dolomitic limestone with intercalations of red sandy limestone beds and black chert lenses; and the upper part consists of alternations of bluish gray limestone, marl, and greenish shale. On the other hand, the Geological Investigation Corps of Taebaegsan Region [9] renamed the Mungog Formation as the Samtaesan Formation, while considering the Yeongheung Formation to rest unconformably on the Samtaesan Formation. Son et al. [27] considered that the Samtaesan Formation cannot be lithologically distinguished from the Machari Formation, and consequently put them together into the Gonggiri Formation. In this study we introduce the revised lithostratigraphic framework of the Mungog Formation based on the stratigraphic succession of dominant lithofacies.

Lithofacies
In this study, four dominant lithofacies are employed to describe the Mungog Formation: i.e., (1) ribbon rock; (2) grainstone to packstone; (3) flat-pebble conglomerate; and (4) marlstone to shale facies.

The ribbon rock facies is characterized by alternating units of calcareous and argillaceous layers, showing a conspicuous banded appearance. Each unit is several millimeters to a few centimeters thick. The lithology of the calcareous layer is variable, ranging from lime mudstone to grainstone. Normal grading, cross-lamination, and bioturbation are often observed within the calcareous layers. A wide structural spectrum exists within the ribbon rock lithofacies, commonly expressed as planar, lenticular, nodular and flaser bedding [6, 21]. The depositional environment of this lithofacies was suggested to be a storm-influenced high intertidal to subtidal setting [21] or a shallow to deep ramp [6].

The grainstone to packstone facies occurs normally as medium- to very thick-bedded and massive to poorly stratified limestone or dolostone. Grains are presumed to be peloids, bioclasts or ooids, although it is difficult to recognize the original grains due to dolomitization [6, 21]. In the lower part of the formation, the chert nodules or layers are intercalated within this facies, which have a good lateral continuity with variable thickness (usually less than 5 cm thick) and often show distinct parallel lamination. This lithofacies was interpreted to have been deposited in a shallow subtidal environment [21] or a shallow ramp [6].

The flat-pebble conglomerate facies is one of the distinctive features of the Mungog Formation and occurs dominantly in the middle to upper parts of the formation. Pebbles are well-rounded, discoidal to spherical in shape and a few millimeters to several centimeters in diameter. They are composed of lime mud or skeletal to non-skeletal grains. The thick flat-pebble conglomerate beds generally have a fairly constant bed thickness, while the thinner ones frequently thin out laterally. Its thickness is normally a few decimeters, but in places it ranges from a few centimeters to two meters. The conglomerates are clast-supported or matrix-supported and poorly sorted; they show a wide variety of intraclast orientation including chaotic, vaguely imbricated, or edgewisely stacked. Paik et al. [21] considered the flat-pebble conglomerates were deposited in a supratidal to subtidal setting, whereas Choi et al. [6] interpreted they were deposited in a middle ramp to basinal environment.

The marlstone to shale facies occurs restrictedly in the upper part of the formation. This facies is greenish gray to dark gray in color and ranges from 20 cm to 2 m in thickness. The unit bed generally starts with massive shale and with increasing micrite content grades into crudely bedded marlstone, although occasionally the reverse relationship is also observed. The crudely bedded marlstone frequently grades into the ribbon rock facies, which is normally truncated by the flat-pebble conglomerate facies. Choi et al. [6] interpreted this facies was deposited in a low-energy deep ramp to basinal environment.

Subdivision of the Mungog Formation
The Mungog Formation is subdivided into four members based on the association of the lithofacies described in the preceding: namely, basal member, ca. 45 m thick; lower member, 30-35 m thick; middle member, 35-60 m thick; and upper member, 50-60 m thick. Figure 2 is a simplified columnar section of the Mungog Formation, showing the stratigraphic occurrences of the lithofacies and fossils.

The basal member is composed mainly of ribbon rock and grainstone to packstone with occasional intercalations of thin flat-pebble conglomerate beds. The lower boundary of the Mungog Formation is easily recognized by the occurrence of planar- to nodular-bedded ribbon rock bed, which overlies light gray to gray massive dolostone of the Wagog Formation. Fossils are confined to the lowermost several-m-thick interval of the basal member; they are mostly of disarticulated trilobites and brachiopods. One of the notable features of the basal member is the occurrence of chert layers or nodules within the grainstone to packstone in the middle part of the member. The upper part of the member comprises dominantly ribbon rock and subordinately grainstone to packstone.

The lower member is recognized by a sequence of gray massive grainstone to packstone which is pervasively dolomitized. The massive dolostone rests on the ribbon rock of the basal member and is in turn overlain by the planar- or nodular-bedded ribbon rock of the middle member. The massive dolostone beds occur also in other members of the Mungog Formation, but they are however much thinner attaining rarely up to several meters in thickness. No fossils were found from the lower member.

The middle member is primarily an alternation of ribbon rock and flat-pebble conglomerate beds. A few grainstone to packstone beds occur in the upper part of the member. The association of ribbon rock with flat-pebble conglomerate facies is not unique to the middle member, as similar associations also occur in the basal and upper members. However, the middle member can be easily distinguished from the basal member in lacking the chert layer-bearing grainstone to packstone and from the upper member in lacking marlstone to shale lithofacies. A trilobite species, *Kainella* sp. cf. *K. euryrachis*, has recently been procured from the lowest ribbob rock bed (ca. 30 cm thick) of the member [12].

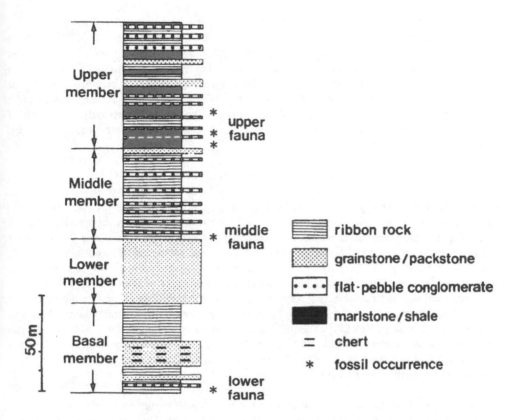

Figure 2. Generalized columnar section of the Mungog Formation, showing the stratigraphic occurrence of the members and trilobite faunas.

The upper member comprises diverse lithofacies including marlstone to shale, flat-pebble conglomerate, grainstone to packstone, and ribbon rock facies. It is characterized by the occurrences of greenish gray marlstone to shale beds, by which this member is distinguished from other members of the formation. The uppermost part of the member is recognized by the occurrence of reddish gray grainstone to packstone which is overlain by the laminated lime mudstone of the Yeongheung Formation. This member yields comparatively diverse and abundant invertebrate fossils including trilobites, brachiopods, ostracods, crinoids, and fossils of uncertain zoological affinities.

BIOSTRATIGRAPHY AND CORRELATION

A total of thirteen trilobite species are identified in the Mungog Formation. The trilobites were procured from the three stratigraphically separated intervals (Fig. 2): i.e., the lowermost several-m-thick interval of the basal member, the lowest bed of the middle member, and the upper member in ascending order. They are herein designated as the lower, middle, and upper fauna, respectively.

Lower Fauna

The lower fauna comprises four trilobite taxa including *Yosimuraspis vulgaris* Kobayashi, 1960, *Jujuyaspis* sp., *Pseudokainella* sp. and Pilekidae gen. and sp. indet (Fig. 3A-E). It consists predominantly of *Yosimuraspis* and subordinately of *Jujuyaspis* and *Pseudokainella* and is supposed to be equivalent to the *Yosimuraspis* Zone [16]. Previously *Asaphellus tomkolensis* was reported in the *Yosimuraspis* Zone [16]. However, no specimens of *Asaphellus* are so far found in the lower fauna. Instead, the association of *Jujuyaspis* and *Pseudokainella* with *Yosimuraspis* seems to be significant, as these genera are known as representative trilobites of early Tremadoc age. Brachiopods occur commonly at certain horizons and include *Obolus*, *Palaeobolus*, *Westonia*, and *Eoorthis*.

To date, no intimate faunal assemblages to the lower fauna of the Mungog Formation are known in other parts of Korea, although it may be correlatable with the *Eoorthis* or *Pseudokainella* Zone of the Duwibong-sequence Joseon Supergroup (Fig. 4, column 2) as the latter two zones have been known to yield *Pseudokainella* [17]. Interestingly, closely comparable faunal assemblages to the lower fauna are better represented in North China. The genus *Yosimuraspis* has been employed as a zonal taxon for the lower Tremadoc sequences of North China [3-5, 8, 31]. The association of *Jujuyaspis* and *Pseudokainella* with *Yosimuraspis* has also been documented in North China [4-5, 8, 31]. In short, the lower fauna of the Mungog Formation is well correlated with the *Yosimuraspis* Assemblage Zone of North China (Fig. 4, column 3). The genera *Yosimuraspis* and *Pseudokainella* have also been reported from the Yangtze Platform [23] and Jiangnan Slope Belt [24] of South China, respectively. Based on these occurrences, the lower fauna may be comparable to part of the *Apatokephalus yanheensis-Songtaia cylindrica* Assemblage Zone of the Yangtze Platform (Fig. 4, column 4) and *Onychopyge-Hysterolenus* Zone of the Jiangnan Slope Belt (Fig. 4, column 5).

The genus *Jujuyaspis* has a short stratigraphic range and wide geographic distribution [1, 28] and thus provides a basis for the precise correlation of the lower fauna with other parts of the world. In Scandinavia, *Jujuyaspis keideli norvegica* occurs within the *Boeckaspis* Zone at a slightly higher horizon than *Dictyonema flabelliforme*. The putative Cambrian-Ordovician boundary has been drawn at the base of the *Boeckaspis* Zone, about three meters below the *Jujuyaspis*-bearing horizon [2]. The occurrence of *Jujuyaspis keideli* has been reported from the *Parabolina argentina* Zone of South America [1, 10]. In North America, *Jujuyaspis* occurs in the *Symphysurina bulbosa* Subzone [7, 20, 28-29].

Middle Fauna

The middle fauna is represented by the sole occurrence of *Kainella* sp. cf. *K. euryrachis* Kobayashi, 1953 (Fig. 3F-H). The genus *Kainella* has been recorded from North America, South America, and Asia. In Korea, *Kainella euryrachis* was known to occur from the *Asaphellus* Zone of the Duwibong-sequence Joseon Supergroup, although the precise stratigraphic position of *Kainella euryrachis* appears to be uncertain [15]. A number of *Kainella* species have been reported from the zones D and E of North America [7] and from the *Kainella meridionalis* Zone of Argentina [10]. Three species of *Kainella* were previously documented in China [8, 25] and Vietnam [15]: however, all of these asiatic species were erected based on cranidia which also appear to lack the typical features of *Kainella*; and without associated pygidia it seems inappropriate to place them to *Kainella*. In summary, the middle fauna can be correlated with the zones D or E of North America and *Kainella meridionalis* Zone of South America. This conclusion is consistent with the occurrence of *Jujuyaspis* in the lower fauna of the Mungog Formation, *Symphysurina bulbosa* Subzone of North America,

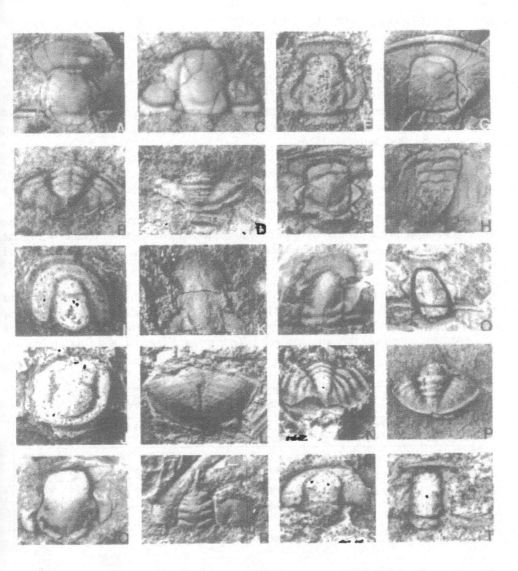

Figure 3. Representative trilobites from the lower, middle and upper faunas of the Mungog Formation, Yeongweol, Korea. Lower fauna: A. *Yosimuraspis vulgaris* Kobayshi, 1960, cranidium, x 1.3; B. *Yosimuraspis vulgaris* Kobayshi, 1960, pygidium, x 3.2; C. *Jujuyaspis* sp., cranidium, x 1.6; D. *Jujuyaspis* sp., pygidium, x 2.7; E. *Pseudokainella* sp., cranidium, x 9.6. Middle fauna: F. *Kainella* sp. cf. *K. euryrachis* Kobayashi, 1953, cranidium, x 2.1; G. *Kainella* sp. cf. *K. euryrachis* Kobayashi, 1953, cranidium, x 0.7; H. *Kainella* sp. cf. *K. euryrachis* Kobayashi, 1953, pygidium, x 1.6. Upper fauna: I. *Micragnostus coreanicus* Kobayashi, 1960, cephalon, x 12; J. *Micragnostus coreanicus* Kobayashi, 1960, pygidium, x 10; K. *Asaphellus* sp., cranidium, x 3.9; L. *Asaphellus* sp., pygidium, x 3.3; M. *Dikelokephalina asiatica* Kobayashi, 1934, cranidium, x 3.2; N. *Dikelokephalina asiatica* Kobayashi, 1934, pygidium, x 4.5; O. *Hystricurus megalops* Kobayashi, 1934, cranidium, x 4.3; P. *Hystricurus megalops* Kobayashi, 1934, pygidium, x 5.7; Q. *Apatokephalus hyotan* Kobayashi, 1953, cranidium, x 6.3; R. *Koraipsis spinus* Kobayashi, 1934, cranidium, x 2.6; S. *Shumardia pellizzarii* Kobayashi, 1934, cranidium, x 16; T. *Hukasawaia cylindrica* Kobayashi, 1953, cranidium, x 16.

and *Parabolina argentina* Zone of South America.

Upper Fauna

The upper fauna consists of eight trilobite species: they are *Micragnostus coreanicus* Kobayashi, 1960, *Shumardia pellizzarii* Kobayashi, 1934, *Apatokephalus hyotan* Kobayashi, 1953, *Hystricurus megalops* Kobayashi, 1934, *Dikelokephalina asiatica* Kobayashi, 1934, *Asaphellus* sp., *Hukasawaia cylindrica* Kobayashi, 1953, and *Koraipsis spinus* Kobayashi, 1934 (Fig. 3I-T). Apart from trilobites, observed are brachiopods, ostracods, crinoids, *Sphenothallus*, *Plumulites*, and *Anatifopsis* [11, 16].

The upper fauna as a whole shows a close resemblance to the Dumugol fauna of the Duwibong-sequence Joseon Supergroup [13-14] in sharing a number of invertebrate taxa between two faunas. However, it should be noted that although the overall similarity between the upper fauna of the Mungog Formation and the Dumugol fauna is apparent, the biostratigraphic zones of the Dumugol Formation are not directly applicable to the Mungog Formation. In other words, the upper fauna collectively comprises elements of the *Asaphellus*, *Protopliomerops*, and *Kayseraspis* Zones of the Dumugol Formation (Fig. 4, column 2)

Comparable faunas to the upper fauna have been well documented from the upper Tremadoc sequences of North China and Australia. The *Asaphellus trinodosus* Zone [5, 31] or *Koraipsis* Zone [8] of the Yehli Formation, North China shares some trilobite genera with the upper fauna of the Mungog Formation: they are *Asaphellus*, *Hystricurus*, *Koraipsis*, and *Dikelokephalina*

	KOREA		NORTH CHINA	SOUTH CHINA		AUSTRALIA	NORTH AMERICA	ARGENTINA
	Yeongweol sequence	Duwibong sequence		Yangtze Platform	Jiangnan Slope Belt	Amadeus Basin		
Arenig	Yeongheung Formation	Kayseraspis Zone		Hunghuayuan Formation	Taohuashi Formation	Horn Valley Siltstone Assem. 3	G	Kayseraspis asaphelloides Zone
Tremadoc	Mungog Formation / Formation upper fauna	Protopliomerops Zone	Asaphellus trinodosus Zone	Acanthograptus - Tungtzuella Zone	S. acutifrons - Asaphopsoides Zone	Assem 2	F	Notopeltis orthometopa Zone
Tremadoc	Mungog Formation / middle fauna	Asaphellus Zone	Euleostegium - Anstokainella Zone	Dactylocephalus letus	A. latilimbatus - T. affinis Zone	Pacoota Sandstone	E / Leiostegium - Kainella Zone	T. tetragonalis - S. minutula Zone
Tremadoc	Mungog Formation / lower fauna	Pseudokainella Zone		Asaphellus inflatus Zone		Pacoota Sandstone	Paraplethopeltis Zone Bellefontia - Xenostegium Zone	Kainella meridionalis Zone
Tremadoc	Mungog Formation / basal member lower fauna	Pseudokainella Zone	Yosimuraspis Zone	A. yanheensis - S. cylindrata Zone	Onychopyge - Hysterolenus Zone	Pacoota Sandstone	Symphysunna Zone	Parabolina argentina Zone
Camb.	Wagog Formation	Eoorthis Zone	Leiostegium Zone	Loushankwan Group	Shenjiawan Formation		Missisquoia Zone	
	1	2	3	4	5	6	7	8

Figure 4. Correlation of the trilobite faunas of the Mungog Formation with coeval ones of other parts of the world. Columns 1-8 are based on the information of the following papers: 1, this study; 2, [13-14]; 3, [3-5]; 4, [23]; 5, [24]; 6, [26]; 7, [7, 20]; 8, [10].

among others (Fig. 4, column 3). The fairly abundant and diverse upper Tremadoc trilobite faunas were reported from the South China [24]. Although their faunal contents are quite different from that of the upper fauna of the Mungog Formation, the occurrence of some cosmopolitan genera such as *Asaphellus*, *Shumardia*, and *Apatokephalus* makes it possible to compare the upper fauna with part of the *Apatokephalus latilimbatus-Taoyuania affinis* or *Shumardia acutifrons-Asaphopsoides* assemblage zones of the Jiangnan Slope Belt (Fig. 4, column 5). An upper Tremadoc trilobite assemblage from the Pacoota Sandstone, Australia [26], designated as Assemblage 2, shows a close similarity to and thus is correlative to the upper fauna of the Mungog Formation (Fig. 4, column 6) in having a fair number of co-occurring genera such as *Asaphellus*, *Apatokephalus*, *Hystricurus*, *Koraipsis*, and *Shumardia*.

Acknowledgments
We are grateful to Zhiqiang Bai (Peking University, China) for his helpful comments during the preparation of this manuscript. The financial support for this study has come from the Basic Science and Research Institute Program, Ministry of Education, Korea, 1996, Project no. BSRI-96-5405.

REFERENCES

1. F.G. Acenolaza and G.F. Acenolaza. The genus *Jujuyaspis* as a world reference fossil for the Cambrian-Orodvician boundary. In: *Global perspectives on Ordovician Geology*. B.D. Webby and J.R. Laurie (Eds.). pp. 81-92. Balkema, Rotterdam (1992).
2. M.G. Bruton, L. Koch and J.E. Repetski. The Naernes section, Oslo Region, Norway: trilobite, graptolite and conodont fossils reviewed, *Geological Magazine* 125, 451-455 (1988).
3. J.Y. Chen, Y.Y. Qian, Y.K. Lin, J.M. Zhang, Z.H. Wang, L.M. Yin and B.D. Erdtmann. *Study on Cambrian-Ordovician boundary strata and its biotas in Dayangcha, Hunjiang, Jilin, China*. China Prospect Publishing House (1985).
4. J.Y. Chen, Y.Y. Qian, J.M. Zhang, Y.K. Lin, L.M. Yin, Z.H. Wang, Z.Z. Wang, J.D. Yang and Y.X. Wang. The recommended Cambrian-Ordovician global boundary stratotype of the Xiaoyangqiao section (Dayangcha, Jilin Province), China, *Geological Magazine* 125, 415-444 (1988).
5. J.Y. Chen, C. Teichert, Z.Y. Zhou, Y.K. Lin, Z.H. Wang and J.T. Xu. Faunal sequences across the Cambrian-Ordovician boundary in northern China and its international correlation, *Geologica et Palaeontologica* 17, 1-15 (1983).
6. Y.S. Choi, J.C. Kim and Y.I. Lee. Subtidal, flat-pebble conglomerates from the Early Ordovician Mungok Formation, Korea: origin and depositional process, *Jour. Geol. Soc. Korea* 29, 15-29 (1993).
7. W.T. Dean. Trilobites from the Survey Peak, Outram and Skoki Formations (Upper Cambrian-Lower Ordovician) at Wilcox Pass, Jasper National Park, Alberta, *Geol. Surv. Canada Bull.* 389, 1-141 (1989).
8. J.Y. Duan, S.L. An and D. Zhao. *Cambrian-Ordovician boundary and its interval biotas, southern Jilin, Northeast China*. Changchun College of Geology, Changchun, China (1986).
9. Geological Investigation Corps of Taebaegsan Region. *Geological maps of Taebaegsan region*. Geological Society of Korea, Seoul (1962).
10. H.J. Harrington and A.F. Leanza. Ordovician trilobites of Argentina, *Univ. Kansas, Dept. Geol., Spec. Publ.* 1, 1-276 (1957)
11. B.K. Kim, C.H. Cheong and D.K. Choi. Occurrences of *Sphenothallus* ("Vermes") from the Mungog and Yobong Formations, Yeongweol area, Korea, *Jour. Geol. Soc. Korea* 26, 454-460 (1990).
12. D.H. Kim and D.K. Choi. *Kainella* (Trilobita, Early Ordovician) from the Mungog Formation of Yeongweol area and its stratigraphic significance, *Jour. Geol. Soc. Korea* 31, 576-582 (1995).
13. K.H. Kim, D.K. Choi and C.Z. Lee. Trilobite biostratigraphy of the Dumugol Formation (Lower Ordovician) of Dongjeom area, Korea, *Jour. Paleont. Soc. Korea* 7, 106-115 (1991).
14. T. Kobayashi. The Cambro-Ordovician formations and faunas of South Chosen. Palaeontology, Part II, Lower Ordovician faunas, *Jour. Fac. Sci., Imp. Univ. Tokyo, Section II*, 3, 1-84 (1934).

15. T. Kobayashi. On the Kainellidae. *Japan. Jour. Geol. Geogr.* **23**, 37-61 (1953).
16. T. Kobayashi. The Cambro-Ordovician formations and faunas of South Korea, Part VI, Palaeontology, V, *Jour. Fac. Sci., Univ. Tokyo, Section II*, **12**, 217-275 (1960).
17. T. Kobayashi. The Cambro-Ordovician formations and faunas of South Korea. Part 10, Stratigraphy of the Chosen Group in Korea and South Manchuria and its relation to the Cambro-Ordovician formations and faunas of other areas. Section A, The Chosen Group of South Korea, *Jour. Fac. Sci., Univ. Tokyo, Section II*, **16**, 1-84 (1966).
18. T. Kobayashi and T. Kimura. A discovery of a few Lower Ordovician graptolites in South Chosen with a brief note on the Ordovician zones in eastern Asia, *Japan. Jour. Geol. Geogr.* **18**, 307-311 (1942).
19. T. Kobayashi, I. Yosimura, Y. Iwaya and T. Hukasawa. The Yokusen geosyncline in the Chosen period. Brief notes on the geologic history of the Yokusen orogenic zone, 1, *Proc. Imp. Acad., Tokyo* **18**, 579-584 (1942).
20. J.D. Loch, J.H. Stitt and J.R. Derby. Cambrian-Ordovician boundary extinctions: implications to revised trilobite and brachiopod data from Mount Wilson, Alberta, Canada, *Jour. Paleont.* **67**, 497-517 (1993).
21. I. S. Paik, K.S. Woo and G.S. Chung. Stratigraphic, sedimentologic and paleontologic investigation of the Paleozoic sedimentary rocks in Yeongweol and Gabsan areas: depositional environments of the Lower Ordovician Mungok Formation in the vicinity of Yeongweol, *Jour. Geol. Soc. Korea* **27**, 357-370 (1991).
22. K.H. Park, D.K. Choi and J.H. Kim. The Mungog Formation (Lower Ordovician) in the northern part of Yeongweol area: lithostratigraphic subdivision and trilobite faunal assemblhges, *Jour. Geol. Soc. Korea* **30**, 168-181 (1991).
23. S.C. Peng. Tremadocian stratigraphy and trilobite fauna of northwestern Hunan. 1. Trilobites from the Nantsinkwan Formation of the Yangtze Platform, *Beringeria* **2**, 1-53 (1990).
24. S.C. Peng. Tremadocian stratigraphy and trilobite fauna of northwestern Hunan. 2. Trilobites from the Penjiazui Formation and the Madaoyu Formation in the Jiangnan Slope Belt, *Beringeria* **2**, 55-171 (1990).
25. S.F. Sheng. The Ordovician trilobites of southwest China, *Acta Paleont. Sinica* **6**, 169-204 (1958).
26. J.H. Shergold. The Pacoota Sandstone, Amadeus Basin, Northern Territory: Stratigraphy and Palaeontology, *Bull. Australian Bur. Min. Res.* **237**, 1-93 (1991).
27. C.M. Son, H.S. Kim, K.H. Paik and M.H. Lee. Geologic structure of Yemi-Yeongweol area, *Jour. Geol. Soc. Korea* **5**, 123-143 (1969).
28. J.H. Stitt and J.F. Miller. *Jujuyaspis borealis* and associated trilobites and conodonts from the Lower Ordovician of Texas and Utah, *Jour. Paleont.* **61**, 112-121 (1987).
29. D. Winston and H. Nicholls. Late Cambrian and Early Ordovician faunas from the Wilberns Formation of central Texas, *Jour. Paleont.* **41**, 66-96 (1967).
30. I. Yosimura. Geology of the Neietsu District, Kogendo, Tyosen (Korea), *Jour. Geol. Soc. Japan* **47**, 112-122 (1940).
31. Z.Y. Zhou and R.A. Fortey. Ordovician trilobites from North and Northeastern China, *Palaeontographica Abt. A*, **192**, 157-210 (1986).
32. Z.Y. Zhou and J.L. Zhang. Uppermost Cambrian and lowest Ordovician trilobites of North and Northeast China. In: *Stratigraphy and paleontology of systemic boundaries in China, Cambrian-Ordovician boundary (2)*. pp. 61-163. Anhui Sci. Tech. Publ. House (1984).

Proc. 30ᵗʰ Int'l. Geol. Congr., Vol. 11, pp. 85-90
Wang Naiwen and J. Remane (Eds)
VSP 1997

Lower Silurian (Llandovery) Rugose Coral Assemblage Zones and their Relation with the Depositional Sequence of Upper Yangtze Region, China*

CHEN JIANQIANG AND HE XINYI

Department of Geology, China University of Geosciences, Beijing, 100083, China

Abstract

The Lower Silurian (Llandovery) rugose coral assemblage zones of Upper Yangtze region are reviewed and the relation between the rugose coral faunal succession and the depositional sequences is discussed. The Lower Silurian rugose corals make up 4 assemblage zones. In ascending order they are: the *Dinophyllum-Rhabdocyclus* zone of Aeronian age in the Xiangshuyuan Formation; the *Stauria-Briantelasma* and *Maikottia-Kodonophyllum* zones in the Leijiatun Formation, extending chronologically from early Telychian to middle Telychian; and the *Gyalophylloides-Idiophyllum* zone in the Ningqiang Formation, corresponding to late Telychian. Five third-order sequences in sequence stratigraphy are recognized from the Lower Silurian of Upper Yangtze region. The rugose coral assemblage zones all occur in the highstand systems tracts of the third-order sequences.

Keywords: Upper Yangtze region, Lower Silurian, rugose corals, assemblage zones, sequence stratigraphy

INTRODUCTION

The work of Wang and He[9, 10], He et al.[5], Wang et al.[12], Chen and He [1] laid the foundation for the study of Lower Silurian (Llandovery) rugose coral assemblage zones of Upper Yangtze region(Table 1). Some researches have also been made on the Lower Silurian sequence stratigraphy in this region[2, 3]. In this paper an attempt is made to clarify the relation between the development stages of rugose corals and the Lower Silurian sequence stratigraphy in Upper Yangtze region, including Guizhou, Sichuan and Shaanxi.

Early and Middle Silurian(Llandoverian-Wenlockian) were the first prosperous stage in the development of rugose corals. Columnariida, Streptelasmatida and Cystiphyllida all reached their acme of flourish. In this period, Cystiphyllida and Streptelasmatida were dominant, with 7 families and 36 genera belonging to Cystiphyllida, and 7 families and 43 genera to Streptelasmatida, numbering about 70% of the total genera. In comparison with Late Ordovician, the Early Silurian rugose coral underwent continuous increase in number of families and genera, with remarkably different constituents. At the end of Ordovician, 54% of rugose coral genera became extinct. The extinction event was probably closely related to global glacial event. In Early and Middle Silurian, the appearance percentage of rugose coral genera reached to 54%, and obvious changes of components took place in Streptelasmatidae and Stauriidae, although only a few genera appeared. Moreover, new genera of Dinophyllidae, Paliphyllidae, Entelophyllidae, Pycnactidae, and Acervularidae and some elements of Zaphrentoidida appeared at first.

---

* Sponsored by "the sequence stratigraphy and sea level change(SSLC)", a key project supported by the State Science and Technology Commission of China and the Ministry of Geology and Mineral Resources.

Table 1. Silurian rugose coral assemblage zones of Upper Yangtze region, China

	Wang, H. Z. & He X. Y. (1981, 1983) [9,10]	He X. Y.et al., (1989) [5]	This Paper
G.F.*	*Micula-Ketophyllum* Assemblage	*Micula-Ketophyllum* Assemblage	*Micula-Ketophyllum* Assemblage Zone
N.F.	*Kyphophyllum-Idiophyllum* Assemblage	*Gyalophylloides-Idiophyllum* Assemblage	*Gyalophylloides-Idiophyllum* Assemblage Zone
U.S.F. / L.F.	*Kodonophyllum-Maikottia* Assemblage	*Kodonophyllum-Maikottia* Assemblage	*Maikottia-Kodonophyllum* Assemblage Zone
			Stauria-Briantelasma Assemblage Zone
L.S.F. / X.F.	*Dinophyllum-Rhabdocyclus* Assemblage	*Dinophyllum-Rhabdocyclus* Assemblage	*Dinophyllum-Rhabdocyclus* Assemblage Zone
G.B.	*Borelasma-Sinkiangolasma* Assemblage	*Borelasma-Grewingkia* Assemblage	*Borelasma-Grewingkia* Assemblage Zone

* G.F.: Guandi Formation; N.F.: Ningqiang Formation; U.S.F.: upper Shiniulan Formation; L.S.F.: lower Shiniulan Formation; L.F.: Leijiatun Formation; X. F.. Xiangshuyuan Formation; G.B.: Guanyinqiao Bed.

Late Silurian (Ludlovian and Pridolian) rugose corals tend to decrease in number on the whole, probably related to the decreasing sea caused by Caledonian orogeny. Among the Late Silurian rugose coral faunas, 6 families including 26 genera belong to Cystiphyllida, 6 families with 19 genera to Streptelasmatida, 13 families with 59 genera to Columnariida, 3 families with 6 genera to Zaphrentoidida, and 2 families with only 2 genera to Caniniida. In Late Silurian, the Streptelasmatidae, Stauriidae, Dinophyllidae and Pycnactidae, all typical in Early Silurian, became extinct, and the Kodonophyllidae in Streptelasmatida and Columnariida were most prosperous, amounting to 30% of the total number of genera. In the Columnariida, Entelophyllidae, Endophyllidae, Spongophyllidae and Grypophyllidae were dominant. Grypophyllidae, Disphyllidae and Eridophyllidae of the Columnariida first emerged . Compared with the Early and Middle Silurian forms, the Late Silurian rugose coral faunas were characterized by a large number of dizonal corals with more complex skeletal structure.

RELATION BETWEEN THE RUGOSE CORAL ZONES AND DEPOSITIONAL SEQUENCES

In Upper Yangtze region, the Lower Silurian rugose corals contain 4 assemblage zones: *Dinophyllum-Rhabdocyclus* assemblage zone of Aeronian age, occurring in the Xiangshuyuan Formation; *Stauria-Briantelasma* and *Maikottia-Kodonophyllum* assemblage zones occurring in the Leijiatun Formation, both of them extending from early Telychian to middle Telychian; and *Gyalophylloides-Idiophyllum* assemblage zone of the Ningqiang Formation, corresponding to late Telychian.

In this region, the Lower Silurian, from Rhuddanian (439Ma)   to Telychian (428Ma), was deposited

Table 2. Lower Silurian rugose coral assemblage zones and sequence stratigraphic framework in Upper Yangtze region, China

SERIES	STAGE	GRAPTOLITE ZONE & AGE (Ma)	FORMATION	BOUNDARY	SYSTEMS TRACT	SEQ	ASSEMBLAGE ZONE
WENLOCK	SHEINWO-ODIAN	427 centrifugus murchisoni	Huixingshao Fm.	SB1	HST	6	
				mfs	TST		
LLANDOVERY	TELYCHIAN	428 grandis spiralis	Xiushan Fm.	SB2	HST	5	Gyalophylloides-Idiophyllum
				mfs	TST		
		grienstoniensis	Rongxi Fm.	SB2	HST	4	
		crispus	Majiaochong Fm.	mfs	TST		
				SB2	HST	3	Maikottia-Kodonophyllum
				mfs	TST		
		turriculatus	Leijiatun Fm.	SB2	HST	2	Stauria-Briantelasma
				mfs	TST		
	AERONIAN FRONIAN	sedgwickii	Xiangshuyuan Fm.	SB2	HST	1	Dinophyllum-Rhabdocyclus
	IDWIAN	convolutus gregarius					
	RHUDDANIAN	cyphus vesiculosus acuminatus persculptus	Longmaxi Fm.	mfs	TST		
ASHGILL	HIRNANTIAN	439 Hirnantia Fauna	Guanyinqiao Bed	SB1	LHST		Borelasma-Grewingkia

continuously and is dominated by thick graptolitic shale, shelly carbonate, and red mudstone, and contains five 3rd order sequences(Table 2), with an average duration of 2 .2 Ma, which constitute a well-defined 2nd order sequence [2, 3]. The first 3rd order sequence consists of Longmaxi Fm and

Xiangshuyuan Fm, extending chronologically from Rhuddanian to Aeronian, with a distinct type-I sequence boundary at the bottom. Basinward, the bottom boundary shows continuous deposition at Tongzi, while landward to the central Guizhou oldland it becomes an overlap   disconformity. Two 3rd order sequences were recognized in the Leijiatun Fm, both characterized by type-II boundaries at the bottom. The Majiaochong Fm (TST) and Rongxi Fm (HST) make up the fourth sequence. The fifth sequence is represented by the lower Member of Xiushan Fm . The five rugose coral assemblage zones mentioned above all occur in the highstand systems tracts of the third-order sequences.

The study of sequence stratigraphy indicates that the rugose coral assemblage zones have a close relationship with the depositional sequences. Because the sea level rose continuously during TST, the sediments are generally characterized by shale and argillaceous siltstone with graptolites, cephalopods,   mixed with a few benthos such as bivalves, brachiopods and corals. The horizon of TST is usually corresponding to BA3-4[7, 8]. The HST is chiefly composed of carbonate and argillaceous limestone, suitable for coral growth, and is corresponding to the position of BA3. Therefore, we regard that the four rugose coral assemblage zones are all confined to the HST of third-order sequences.

EARLY SILURIAN RUGOSE CORAL ASSEMBLAGE ZONES OF UPPER YANGTZE REGION

The Lower Silurian (Llandovery)   of northeastern Guizhou may be subdivided into Longmaxi, Xiangshuyuan, Leijiatun, Majiaochong, and Rongxi Formations, in addition to the lower Member of Xiushan Formation (Table 2) in ascending order, and includes 4 rugose coral assemblage zones.

Dinophyllum-Rhabdocyclus assemblage zone
The Longmaxi Formation, approximately corresponding to Rhuddanian, consists largely of shales with graptolites. The earliest known coral-bearing horizon in northeastern Guizhou is the Xiangshuyuan Formation of Aeronian age. The *Dinophyllum-Rhabdocyclus* assemblage zone occurs in the Xiangshuyuan Formation (Table 3) or the lower part of the Shiniulan Formation. Solitary forms are dominant (Table 2, 3), including Cystiphyllida: *Rhabdocyclus, Tryplasma, Aphyllum, Cantrillia, Cystiphyllum, Cysticonophyllum, Microplasma, Protocystiphyllum, Rhizophyllum, Gyalophyllum, Hedstroemophyllum, Dentilasma;* Streptelasmatida: *Crassilasma, Axolasma, Paramplexoides, Yangziphyllum, Brachyelasma, Leolasma, Dinophyllum, Kodonophyllum, Pycnactis, Holophragma, Onychophyllum, Pseudophaulactis;* and Columnariida: *Stauria, Ceriaster, Palaeophyllum, Paraceriaster, Eostauria, Parastauria, Zelophyllum, Amplexoides, Pilophyllia.* The study of sequence stratigraphy shows that this assemblage zone occurs within the HST of the first third-order sequence.

Stauria-Briantelasma assemblage zone and Maikottia- Kodonophyllum assemblage zone
Two rugose coral assemblage zones may be distinguished in the Leijiatun Formation or the upper part of the Shiniulan Formation, the *Stauria-Briantelasma* assemblage zone below and *Maikottia-Kodonophyllum* assemblage zone above. The Leijiatun Formation includes two third-order sequences, the second and the third. The *Stauria-Briantelasma* assemblage zone exists in the HST of the second sequence, in the lower part of the Leijiatun Formation; while the *Maikottia-Kodonophyllum* assemblage zone in the HST of the third sequence, the upper part of the Leijiatun Formation.

The *Stauria-Briantelasma* assemblage zone contains Cystiphyllida: *Aphyllum, Microplasma, Gyalophyllum, Cystiphyllum, Cysticonophyllum, Tryplasma, Tabularia, Rhizophyllum;* Streptelasmatida: *Pycnactis, Briantelasma, Pseudophaulactis, Crassilasma, Holophragma, Yangziphyllum, Kodonophyllum;* and Columnariida: *Stauria, Ceriaster, Paraceriaster, Eostauria,*

Table 3. Taxonomic composition of the Lower Silurian (Llandovery) rugose coral assemblage zones in Upper Yangtze region, China

Series	Assemblage zone	Cystiphyllida	Streptelasmatida	Columnariida	Zaphrentoidida
Wenlock	*Micula-Ketophyllum*	8*	2	3	
Llandovery	*Gyalophylloides-Idiophyllum*	15	5	14	
Llandovery	*Maikottia-Kodonophyllum*	12	6	9	
Llandovery	*Stauria-Briantelasma*	8	7	9	
Llandovery	*Dinophyllum-Rhabdocyclus*	13	13	11	2
Ashgill	*Borelasma-Grewingkia*	2	14		

* 8: number of genera

Amplexoides,Pilophyllia, Synamplexoides, Palaeophyllum. This assemblage zone is similar to the *Dinophyllum-Rhabdocyclus* assemblage zone at genus level, but different at species level and with a large number of forms of Stauriidae as the dominant constituents.

The *Maikottia-Kodonophyllum* assemblage zone includes Cystiphyllida: *Maikottia, Tabularia, Tryplasma, Cantrillia, Hedstroemophyllum, Microplasma, Cystiphyllum, Dentilasma, Ketophyllum, Calceicystiphyllum, Zonocystiphyllum, Dentilasma;* Streptelasmatida: *Kodonophyllum, Brachyelasma, Leolasma, Pycnactis, Crassilasma, Pseudophaulactis;* and Columnariida: *Amplexoides, Pilophyllia, Ceriaster, Stauria, Paraceriaster, Palaeophyllum, Entelophyllum, Zelophyllum, Strombodes. Maikottia* is a massive coral with septa consisting of composite rhabdacanthes, represents a more evolutionary form, and is known from the HST of the third sequence in the Leijiatun and Benzhuang of Shiqian as well as Guangyinqiao, Qijiang, Sichuan. The elements of Stauriidae in this assemblage zone are the more evolutionary forms with dissepimentarium and columellae, such as *Stauria normola, Ceriaster columllatus.* Moreover, dizonal Columnariida, including *Strombodes* and *Entelophyllum,* appears. All these indicate that this assemblage zone is more evolutionary than the two lower assemblage zones.

Only a few rugose corals are known from the Yangpowan Formation and equivalent beds of Ningqiang, Shaanxi and the Modaoya Formation of Guangyuan, Sichuan. They are Cystiphyllida: *Ketophyllum, Tryplasma, Rhizophyllum, Ketophylloides, Aphyllum, Rhabdocyclus, Hedstroemophylum, Cystiphyllum, Tabularia;* Streptelasmatida: *Crassilasma, Pycnactis, Dinophyllum, Kodonophyllum;* Columnariida: *Amplexoides, Pilophyllia, Strombode, Lamprophyllum, Zelophyllum, Ceriaster, Paraceriaster, Entelophyllum, Ptychophyllum, Kyphophyllum.* Dizonal corals are dominant in this fauna, which permit to correlate the fauna to the *Maikottia-Kodonophyllum* assemblage zone of the upper part of the Leijiatun Formation.

Gyalophylloides-Idiophyllum assemblage zone
This assemblage zone occurs in the lower part of the Ningqiang Formation of late Telychian age in southern Shaanxi, corresponding to the upper part of the Lower Number of the Xiushan Formation. Wang and He[9, 10] proposed the *Kyphophyllum-Idiophyllum* assemblage, and Wang et al. [11] established the *Gyalophylloides-Idiophyllum* assemblage zone. The *Gyalophylloides-Idiophyllum* assemblage zone is followed here. This assemblage zone is very high in genera and species

90

diversity, containing a large number of elements of Columnariida and dizonal corals, and seems to be more evolutionary than the coral fauna in the Leijiatun Formation . The known genera are Cystiphyllida: *Gyalophylloides* , *Rhizophyllum, Tryplasma, Tabularia, Hedstroemophyllum, Pseudamplexus, Chonophyllum, Cystiphyllum, Aphyllum, Cystilasma, Ketophyllum, Cysticonophyllum, Diplochone, Ketophylloides, Rhabdocyclus;* Streptelasmatida: *Pseudophaulactis, Crassilasma, Streptelasma, Neobrachyelasma, Palaeocyathus;* Columnariida: *Idiophyllum, Strombodes, Kyphophyllum, Shensiphyllum, Pilophyllia, Amsdenoides, Amplexoides, Zelophyllum, Cyathactis, Lamprophyllum, Lindstroemophyllum, Micula, Miculiella, Ptychophyllum, Ningqiangophyllum.* Jin Chuntai et al. [6] proposed three genera, *Dongquanbaophyllum, Palaeoaulina* and *Paraphillipsastraea.* Here *Dongquanbaophyllum* is regarded as a synonym of *Ningqiangophyllum,* and *Palaeoaulina* as a synonym of *Idiophyllum* [4]. The genus *Paraphillipsastraea* is temporary not used here because its photographs are not clear.

Acknowledgments

We would like to express our cordial thanks to Professor Wang Hongzhen for his direction and encouragement. We are also thankful to Professor Wang Xunlian, Professor Shi Xiaoying and Ms Wang Hua for help in various ways.

REFERENCES

1. Chen Jianqiang and He Xinyi. Ordovician and Silurian rugose coral   faunas and   assemblage zones in Yangtze region, China, *Earth Science.* (in press) (in Chinese with English abstract).
2. Chen Jianqiang, Li Zhiming, Gong Shuyun, Li Quanguo and Su Wenbo, Early Silurian sequence stratigraphy of north-eastern Guizhou, China. *30th International Geological   Congress, Abstracts.* **3.** 32 (1996).
3. Chen Jianqiang, Li Zhiming, Gong Shuyun, Li Quanguo and Su Wenbo. Early Silurian sequence stratigraphy of north-eastern Guizhou, China. *Earth Science.* (in press) (in Chinese with English abstract).
4. He Xinyi and Chen Jianqiang. On the Genus *Idiophyllum* . *Acta Palaeontologica Sinica,***25.** 525-530. (1986) (in Chinese with English abstract).
5.   He Xinyi, Li Zhiming and Chen jianqiang. Ordovician and Silurian rugose coral faunas of China. In: Wang Hongzhen, He Xinyi, Chen Jianqiang and others, (eds.), *Classification, Evolution and Biogeography of the Palaeozoic corals of China.* chapter 11, pp.226-238. Science Press, Beijing, China. (1989) (In Chinese with an English summary).
6.   Jin Chuntai, Ye Shaohua, Chen Jirong, Qian Yongzhen and Yi Yongen. The   Silurian   system   in Guangyuan, Sichuan and Ningqiang, Shaanxi. Chengdu University of Science and Technology, Chengdu. (1992) (in Chinese with English abstract).
7.   M. E. Johnson, Rong Jiayu and Yang Xuechang. International correlation by sea level events in the Early Silurian of North America and China (Yangtze Platform). *Geol. Soc. Amer. Bull.* **96.** 1384- 1397 (1985).
8.   Rong Jiayu, Markes E. Johnson and Yang Xuechang. Early Silurian (Llandovery) sea level changes in the Upper Yangtze region of Central and Southwestern China. Acta Palaeontologica Sinica, **23.** 673-693. (1984) (in Chinese with English abstract).
9.   Wang Hongzhen and He Xinyi. Silurian rugose assemblages and palaeobiogeography of China. *Geol. Soc. Am. Spec. Pap.* **187.** 55-63 (1981).
10. Wang Hongzhen and He Xinyi. Silurian rugose assemblages and palaeobiogeography of China. In:*Palaeobiogeographic Provinces of China.* pp. 32-42. Science Press, Beijing (1983) (in Chinese).
11. Wang Hongzhen, He Xinyi, Li Yaoxi, Li Zhiming, and Chen Jianqiang. Introduction. In: Wang Hongzhen, He Xinyi, Chen Jianqiang and others, (eds.), *Classification, Evolution and Biogeography of the Palaeozoiccorals of China.* chapter 1, pp.1-5. Science Press, Beijing, China. (1989) (In Chinese with an English summary).
12   Wang Hongzhen, Wang Xunlian and Chen Jianqiang. Evolutional stages and biogeography of the rugose corals.   In: Wang Hongzhen, He Xinyi, Chen Jianqiang and others, (eds.), *Classification, Evolution and Biogeography of the Palaeozoic corals of China.*chapter 10, pp.175-225. Science Press, Beijing, China. (1989) (In Chinese with an English summary).

Proc 30ᵗʰ Int'l. Geol. Congr., Vol. 11 , pp. 91-97
Wang Naiwen and J. Remane (Eds)
© VSP 1997

TETHYS—AN ARCHIPELAGIC OCEAN MODEL

YIN Hongfu
(China University of Geosciences, Wuhan, Hubei, China, 43004)

Abstract

Three models for Tethys are discussed. Based on recent data from China and adjacent areas, it is concluded that an archipelagic ocean model for Tethys accords with paleomagnetic and biogeographic data that Eurasia and Gondwana were widely apart, and also accords with sedimentary data that deep ocean deposits were few, because the vast distance were filled by islands.

So far there are three models for Tethys. The shallow seaway model holds that, judging from the abundant benthic biotas, carbonate and shallow clastic facies found from Alps to Himalaya, a vast and deep Tethys ocean separating Eurasia and Gondwana never existed in geological history; instead it was composed of epicontinental seas and relatively deeper seaways. To explain the wide distance between southern and northern supercontinents necessitated by paleomagnetic, palaeoclimatic and biogeographic desparation, this view resorts partly to the theory of earth's expansion in time upheld by Carey (1975) and Owen (1992), by which the above distance was reduced to accord with the seaway hypothesis, and the palaeomagnetic data were regarded by them as unreliable. However, according to recent geophysical and geological data, the expansion rate of 20 or even 40 per cent is exaggerated. If we take a moderate view of expanding earth with 20% enlargement of radii since Archaic, length of the earth's radii at Palaeo-Tethys interval (ca. 300 Ma ago) will be about 95% of the modern radii, and the surface dimension at that age, using the equation $S=4\pi r^2$, would be $0.95 \times 0.95 = 0.9$ of the modern surface dimension, which does not help much for the narrow seaway model. Also the palaeomagnetic data have been acumulated to such an extent that it seems unavoidable to place the southern and northern super- continents widely apart.

The second or most popular model for Palaeo-Tethys is a wedge-shape ocean tipping at its western end at western Alps while opening eastward into Panthalassa or Palaeo-Pacific. Figures of Scotese and McKerrow (1990), for example, show a relatively `clean' ocean. This model has synthesized modern data of palaeomagnetics, paleoclimate and palaeo- biogeography. However it can not explain the question why there is so few evidences of ocean crust found in this vast region. The following facts now quite convinceably established are against such a 'clean' model.

1. Most of the biota and facies found along Palaeo-Tethys are of shallow water type, although more and more flysch, radiolarites and other deep water deposits have been reported recently, the bulk of sediments remains belonging to continent crust. Evidences of ocean crust are relatively few.

2. Most of the ophiolites found along Palaeo-Tethys belong to types of marginal and back-arc seas. Typical oceanic ophiolites are few.

These facts have cast doubts to the truth of the second model and led to recommendation of the present model--the archipelagic model of Tethys. It is supported by recently accumulated data from China. The micro-ocean between Yangtze and North China Plates used to be regarded as 'clean', with northern and southern Qinling belonging to continental margin of the two plates respectively. However, recent research has revealed that there was an intermediate row of blocks between the two plates, consisting of Dabie, Zhen'an-Xichuan, western Qinling and constituting the Qinling Microplate (Yin et Huang, 1995) (Fig. 1), bordered by the Shangxian--Danfeng suture to the north and the Mianxian-Lueyang suture to the south. Between the Qinling Microplate and North China, there were during Latest Proterozoic (Sinian) -- Early Palaeozoic at least two rows of island arcs: the southern arc composed of the Qinling Group and the northern one composed of Kuanping Group. Moreover, islands existed in the passive marginal seas and epicontinental seas of Yangtze. Thus, this area was not a 'clean' micro-ocean but an archipelagic micro-ocean.

South China also used to be regarded as more or less integrate except for the Hainan Island and western Yunnan which have long been considered as pro-Gondwanan. However, it has been discovered recently that Early Palaeozoic and Carboniferous-Permian microfossils exist in cherts associated with 'Proterozoic' ophiolites and intercalated within 'Proterozoic' flysch (the Shuangqiaoshan or Banxi Group) in Yiyang of Zhejiang-Jiangxi border, that Late Palaeozoic melanges and radiolarians in cherts extend from the Qingzhou-Fangcheng area to Yuling, Yunfu and further to Yinde, that large-scale marine basalt exists in western Guangxi which was previously regarded as diabase sill-swarm. These discoveries suggest existence of micro-oceans or deep seas separating South China into blocks. These plus sedimentary and palaeomagnetic data have resulted in a much more fragmentary pattern of

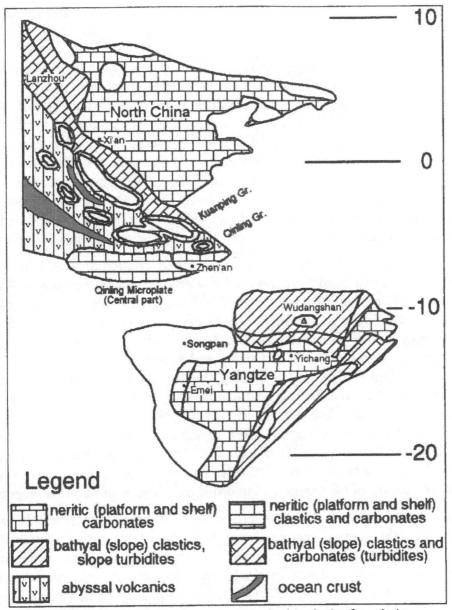

Fig. 1 The Qinling micro-ocean during Late Cambrian

South China in geological history.

The Tibet-Qinghai Plateau has long been recognized as composed of microplates and island arcs demarcated by sutures, such as the northern and southern Qiangtang, Tanggula, Gangdise, Qamdo, Lanping-Simao, Baoshan,

Tengchong etc. The mosaic can be more complicated judging from recent compilation by Pan et al. (1996).

To the northern border of China, the Palaeo-Asian Ocean showed similar archipelagic pattern during its maximum development in Early Palaeozoic. So is the southeastern Asia in Mesozoic according to many authors.

Fig. 2 shows the distribution of continental blocks (microplates, rifting-away blocks, etc.) which displays an archipelagic view in eastern Tethys.

The plate tectonics of this model should have certain distinctiveness, and should not mimicry models derived from the Atlantic and Pacific. The suturing process of Yangtze and North China shows following characteristics different from classic plate collision: 1) North China, Qinling and Yangtze moved along the same (northward) direction but with different speeds and directions of rotation, and finally led to closure of micro-ocean and plate suturing. The motional energy ($1/2mv^2$) produced by such process is much smaller than that of head-on collision between two plates, because in the latter case v is the sum of velocities of the two plates, whereas under the situation of Qinling v is the difference of two blocks or plates. 2). The suturing process between Yangtze and North China underwent following steps: first, the two island arcs of Qinling and Kuanping Groups accreted to North China margin; next, the Qinling Microplate contacted the newly accreted arcs; then the Lueyang-Mianxian (maybe rift belt during Early Palaeozoic) subtracted and closed, and Qinling contacted Yangtze; finally Yangtze and North China Plates sutured together in a scissor-type from east to west. The collision was not performed at one stroke, but stepwise. Each individual step may be merging of islands, block accretion onto continental margin, or closure of rift trough and micro-ocean. The m of motional energy ($1/2mv^2$) produced during each micro-scale collision is thus small. If we imagine that the sum of motional energy produced by collision between Yangtze and North China disintegrated into three or four parts of micro-scale collision between islands or block accretion, that the velocity of each micro-scale collision was the difference rather than the sum of velocities of two colliding blocks, and that the sum of mass of the two colliding blocks was small , then the result would be that a saltation fell apart into a series of gradations. Each step of this process produced little energy, definitely incomparable with that from 'ordinary' plate collision, thus comprehensibly did not caused an orogeny. This is called 'soft collision'. It makes one characteristic of the plate movement in the archipelagic Tethys.

Fig. 2 Distribution of blocks in eastern Tethys region
1. Siberia; 2. Mongolia-Ergun; 3. Tarim; 4. North China (Sino-Korea)
5. Qiangtang; 6. Lhasa; 7. Yangtze; 8. Lower Yangtze; 9. Cathaysia
10. Indosinia; 11. Sibumasu; 12. India

The second distinctiveness is that in many cases orogeny did not immediately follow block collision, but may delay up to 100 Ma or more. Take Qinling for example, the first 'soft collision' of Yangtze and North China occurred at Siluran-Devonian transition, but the northern continental margin of Yangtze did not experience orogeny until Middle Triassic. This is because most Tethyan blocks taking part in collision were moving unidirectionally (northward) due to Gondwanan dispersal. According to Newton's Law of action and reaction, each collision caused by unidirectional movements with unequal velocities can speed up the northward shift of the foregoing block, meanwhile the reaction can slow down the northward migration of the back one. Thus after each collision will follow a process of extension, or increase of distance between the two blocks. This is the mechanics of multi-cyclic activation characteristic in Chinese geology. Theoretically an orogeny can occur only when the forgoing block collided to a fixed plate and stopped, and the northward drifting back block collided with the foregoing one and proceeded to perform intracontinental underthrust, thus producing compression. In the case of Qinling, timing of the stopping of North China may be regarded as in Late Permian when North China finally collided with southern margin of Siberia. After that Yangtze proceeded to drift northward and therefore the orogeny did not occur until Middle Triassic. This differs from the collision between India Subcontinent and Asia, in which case Asia was a relatively fixed macro-continent. After its collision with Asia at ca. 80 Ma ago India proceeded to perform intracontinental subduction and began to form the Himalayas at ca. 20 Ma ago, with a time interval of 60 Ma between collision and orogeny. The case comparable with Qinling is the Bangong Cuo-Dongqiao-Dingqing Jurassic micro-ocean in northern Tibet. The timing of its closure and suturing is end-Jurassic—earliest Cretaceous (ca. 150 Ma ago), whereas the timing of its uplift or the formation of northern Tibet Plateau is estimated to be post-Neogene (ca. 23 Ma ago). The time difference is more than 100 Ma. Here again the process refers to two micro-plates, namely Qiangtang-Tanggula and Gangdise-Lhasa, similar to the situation in Qinling.

To sum up, the third or archipelagic model seem more appropriate for Tethys and its predecessors, especially the eastern Tethys. Palaeotethys, Mesotethys, Neotethys and modern Indian Ocean are all formed during the general process of Gondwana dispersal and Asia accretion, through extension and rift in the back of disintegrated belt of blocks such as Tethsides and Cimmerides (*sensu* Sengor). The modern Indian Ocean is formed through extension and rifting away of the belt of blocks including Arabian Penisular, Indian Subcontinent and Australia from Africa and Antarctica, and in this sense could be called Modern Tethys. There has been persistently a plenitude of disintegrated blocks within these ancient and modern oceans, from Palaeotethys to Indian Ocean; the foregoing (northerly) blocks were accreting to Eurasia, while the back (southerly) ones were rifting away from Gondwana. Therefore, they have been persistently archipelagic oceans, different from the Atlantic and Pacific Oceans. This accords with paleo-magnetic and biogeographic data that Eurasia and Gondwana were widely apart, and also accords with sedimentary data that deep ocean deposits were few, because the vast distance were filled by islands. Researches on the plate movement process and mechanics of such archipelagic micro-oceans may help to comprehend many plate phenomena in geological history and to predict analogus evolutions such as that of South China Sea—Indonesia Archipelago micro-ocean.

References

Carey, S.W., 1975, The expanding earth-an essay review. Earth Science Review, Vol. 11, No. 2.

Owen, H.G., 1992, Has the earth increased in size? In Chattergee S. and N. Hotton (eds.), New concepts in global tectonics. Lubbock, Texas Tech. Univ. Press, 450 pp.

Pan Guitang, Chen Zhiliang, Li Xingzhen et al., 1996, Models for the evolution of the polyarc-basin systems in eastern Tethys. Sedimentary Facies and Palaeogeography, 16(2): 52-65.

Scotese, C.R., and McKerrow, W.S., 1990, Revised world maps and introduction, Palaeozoic palaeogeography and biogeography. Memoir of the Geological Society of London, 12, 1-24.

Yin Hongfu and Huang Dinghua, 1996, Early papaeozoic evolution of the Zhen'an-Xichuan Block and the small Qinling Multi-island Ocean Basin. Acta Geologica Sinica, 9(1):1-15.

Proc. 30ᵗʰ Int'l. Geol. Congr., Vol. 11 , pp. 99-113
Wang Naiwen and J. Remane (Eds)
© VSP 1997

An Integrated Chronostratigraphic Scheme for the Permian System

JIN YU-GAN
Laboratory of Palaeobiology and Stratigraphy, Nanjing Institute of Geology & Palaeontology, Chinese Academy of Sciences, Nanjing 210008, China

Abstract

International agreement has been achieved on the names of series and stages into which the Permian System should be divided and on the levels for the basal boundaries of these subdivisions. This scheme incorporates with three excellent reference successions: The Cisuralian Series of the Urals, the Guadalupian Series of SW USA and the Lopingian Series of South China respectively for the Lower Permian, the lower and upper series of the Upper Permian. The scheme will serve as a working template for the Permian Subcommission in defining the GSSPs for intra-systemic boundaries and also enables us to correlate the Permian marine sequences world wide in a higher resolution manner than the previous.

Keywords: Chronostratigraphy, Permian, Cisuralian, Kungurian, Guadalupian, Lopingian

INTRODUCTION

The Permian Subcommission has been attempting for two decades to establish a global time scale for the Permian System and the GSSPs for its initial boundary and intra-systemic boundaries. This has been widely known as a difficult task since that the sequence above the Artinskian of the traditional standard sections in the Urals is composed of facies not favorable for defining the GSSPs of relevant boundaries, and that the available stratigraphic data of other regions have long been inadequate in establishing a precise and reliable correlation of post-Artinskian successions in strongly differentiated biogeographic regions. We are pleased to see that the conclusion of such a long story to integrate the suitable marine successions into a single Permian chronostratigraphic scheme seems now to be approaching.

A generally agreed scheme has been achieved through the development of several drafts during last two decades.

Composite Schemes of Regional Stages
Attempts to build up a composite scheme based on marine sequences as a substitute of the traditional standard were launched in the 1960s (10). Subsequently, composite successions integrating various regional stages were proposed based mainly on the interpretation of evolutionary succession of the regionally limited ammonoid, fusulinid and brachiopod faunas (9, 26, 44, 47). However, none of these has gained an

overwhelming acceptance because it is often impossible to prove the objective stratigraphic superposition of neighboring stages. Communications through the newsletters of the Subcommission (Permophiles) led the members to realize that it is better to integrate the reference successions from a minimum number of type regions into a standard succession. The scheme suggested by Waterhouse (47) reflected the momentum brought about by the Subcommission.

Composite Successions of Regional Series
In 1985, an international committee of the Subcommission was organized. The regional correlation charts worked out by various national working groups in 1987, and discussed in Beijing, showed a considerable consistency in key stratigraphic levels of the marine successions above the Kungurian. In 1988, on behalf of the Subcommission, we recommended Harland and his collaborators to replace the Kazanian and Tatarian stages with the Guadalupian and Lopingian Subseries. They accepted it and considered the introduction of the Lopingian Subseries was a notable novel feature of their new version of the global time scale (14).

In Perm, Russia, 1991, the succession of Asselian, Sakmarian and Artinskian was further documented as a potentially qualified international standard (4), and the Guadalupian Series was formally proposed as a global standard (11). International working groups were organized for erecting the chronostratigraphic successions corresponding respectively with the Lopingian, the Pre-Artinskian and that between the Guadalupian and Artinskian.

An Operational Scheme
Leaving aside the widely different usage of the names Early, Middle and Late Permian, an operational scheme incorporating three most promising reference successions was proposed as a working template for the Permian Subcommission in 1994 (17). Updated fossil zones of conodonts, ammonoids and fusulinids were selected for the scheme. It was following a number of comments, which indicated that among four series recommended, the Uralian and the Lopingian Series (2, 7, 12, 19, 22, 27) are generally acceptable as they are privileged by their historic priority, relatively complete succession and extensive studies. Two middle Permian series, particularly, the Chihsian \ the Cathedralian Series in the proposed scheme can not be defined precisely because of the uncertainty in correlation between the Tethyan and the North American successions. More data are desirable for establishing this unit.

Revision and Decision
During the International Guadalupian Symposium II (IGS II)) in Alpine, USA, 1996, the Working Group on Post-Artinskian Series suggested to retain the name of the Kungurian in the scheme and to define its initiation at the base of the *Neostreptognathodus pnevi - N. exsculptus* Zone with the understanding that this stage should be established based on full marine successions in North America or in Tethys because of the paucity of more useful biostratigraphic groups such as the conodonts, ammonoids and fusulinids of the Kungurian in its eponymous area. This proposal, together with the usage of the Lopingian Series and its two subdivisions, namely, the Wuchiapingian and the Changhsingian Stage for the uppermost series of the Permian

System, and the Guadalupian Series and its subdivisions, namely, the Roadian, the Wordian and the Capitanian Stage for the lower series of the Upper Permian were passed by voting at the Subcommission's meeting .

This agreement was further supported by a postal ballot. The result of this ballot is decisive in favor of the revised chronostratigraphic scheme for the Permian System because nearly all Titular Members of the Permian Subcommission agree with the usage of the Cisuralian, Guadalupian and Lopingian Series and their component stages respectively for the Lower Permian, the lower and upper series of the Upper Permian (20).

Carboniferous-Permian Boundary

With regard to the basal boundary of this system, the International Working Group on the Carboniferous \ Permian boundary was re-organized in Beijing, 1987. Two years later, the potential stratigraphic levels for the C\P boundary were narrowed down to two horizons at a meeting of the WGCP during the 28th IGC in Washington D. C. A proposal to define the Carboniferous-Permian boundary at Aidaralash Creek, Northern Kazakhstan was passed in the Subcommission and the Commission on Stratigraphy in 1995, and now has been ratified by the executive board of IUGS. It is marked by the first appearance of *Streptognathodus isolatus*. The boundary level is located within the Bed 19, slightly below the boundary between the *Shumardites-Vidrioceras* and the *Svetlanoceras-Juresanites* Genozone, and approximately coincides with the base of *Sphaeroschwagerina vulgaris - S. fusiformis* Zone (6).

Series		Stages	Basal conodont zone	Jin et al., 1994		Harland et al., 1989		Waterhouse, 1982	
P E R M I A N	Lopingian	Changhsingian	*Clarkina subcarinata*	Lopingian	Changhsingian	Zechstein	Lopingian	Changhsingian	Lopingian	Changhsingian
		Wuchiapingian	*Clarkina postbitteri*		Dzhulfian (Wuchiapingian)		Lungtanian		Lungtanian
	Guadalupian	Capitanian	*Jinogondolella postaserrata*	Guadalupian (Cathedralian)	Capitanian	Guadalupian	Capitanian	Guadalupian	Capitanian
		Wordian	*Jinogondolella aserrata*		Wordian		Wordian		Wordian
		Roadian	*Jinogondolella nankingensis*		Roadian		Ufimian		Roadian
	Cisuralian	Kungurian	*Neostreptognathodus pnevi- N. exsculptus*	Chihsian (Cathedralian)	Kubergandinian Bolorian	Rotliegendes	Kungurian	Cisuralian	
		Artinskian	*Sweetognathus whitei- Mesogondolella bisselli*	Uralian	Artinskian		Artinskian		Baigendzinian
		Sakmarian	*Streptognathodus postfusus*		Sakmarian		Sakmarian		Sakmarian
		Asselian	*Streptognathodus isolatus*		Asselian		Asselian		Asselian

Figure 1. Development of Permian chronostratigraphic scales
This chart is designed to show the succession of chronostratigraphic units in selected scales rather than the correlation between them.

GENERAL APPROACH

The proposed scheme (17, 19, 20) has evoked diverse comments from various aspects as we expected, for there are many different views on the approach to the chronostratigraphic classification, nomenclature and procedure, and Permian stratigraphers who essentially agree with the approach are diverged in selection of boundary levels and their stratotypes. The problem becomes even more complicated since Permian stratigraphers in Boreal and Gondwanan areas will be confronted with the difficulty in adopting the chronostratigraphic subdivisions based on the successions in Tethyan areas and North America , and the non-conodont experts will have the similar problem with the standard successions based on evolving conodonts.

Comments dealing with the general aspects of the approach focus on the question whether to define the chronostratigraphic units based on the successions of highly diversified marine faunas or on the traditional standard. Relevant points of view have been presented in "The International Stratigraphic Guide" (41), and in "The Guidelines for Establishment of Global Chronostratigraphic Standards by the ICS" (37). However, a brief explanation on the following points is necessary.

Necessity of An Integrated Scheme

Concerns have been expressed that the approval of the proposed scheme will entail rejection of the traditional standard in Urals which has been used for over 150 years (Ganelin et al., a letter to Chairman of Permian Subcommission, 1996). These concerns are understandable but are in reality, unwarranted. The Permian succession of the Urals will continually serve as traditional standard. The founding of an integrated chronostratigraphic scheme will not devalue the inherent significance of traditional standard in intercontinental correlation.

Nevertheless, all well known regional successions, including the traditional standard have been extrapolated into some form of local zonal arrangement but all lack global compatibility. Consequently, none has achieved chronostratigraphic status in accordance with the guidelines published by the ICS. Although used for more than a century, the classic stages of traditional standard have never served effectively as precisely defined chronostratigraphic units. Those above the Artinskian are particularly inadequate. In most cases, the traditional standard was cited in correlation parallel to other major regional standards, the Tethyan, North American, Gondwanan, and West European. The necessity is obvious to establish a Permian chronostratigraphic scheme defined on the successions with better biostratigraphic precision and correlation potential.

It is generally agreed that a precise and reliable chronostratigraphic scheme can be defined by the GSSPs for the systemic and intra-systemic boundaries. Those who agree with this operational framework and the requirements for a GSSP will find that there is little potential for erecting the GSSPs for the chronostratigraphic subdivisions above the Artinskian in the traditional standard succession. In this case, we have no choice but to consider the potential candidates in Tethyan areas and North America. Analysis of evolutionary changes can be made on the high diversified faunas in palaeoequatorial areas much more in detail than on relatively impoverished faunas in the high palaeolatitudinal areas. Therefore, the relevant boundaries can be defined precisely.

Favorable Facies and Key Fossil Group
By the Subcommission, it has been recognized in general, that the open marine facies forms a better basis for the biostratigraphic precision needed for international correlation. Moreover, the evolutionary clines of conodonts can provide a precise boundary level of great correlation potential. This is not to imply that there are not facies rich in other groups, for example fusulinids, ammonoids, spores and pollens, which are very important for correlation, but it is normally easier to correlate into such sections secondarily from primary sections in the open marine facies. Disapproval was emphasized in the comments on the facies and key fossils favored by the Subcommission in defining the proposed scheme (42). However, the achievement of the ICS bodies during last decades are highly convincing. Nearly all ratified GSSPs of the Paleozoic are defined using the evolution of conodonts. Practical experience of the Permian Subcommission have again demonstrated that such a trend can be continued with the definition of Permian chronostratigraphic units. For instance, the succession of conodont zones can be precisely identified in Cisuralian sequences of Urals, Southwest USA and South China. Of course, many parts of the conodont succession need urgently to be elaborated, classification of the leading taxa of conodonts should be clarified based on well represented samples, correlation between the successions of conodonts and other leading fossils should be erected as much as we can, and so forth. Nevertheless, all this implies merely that it is necessary to invest more in this framework.

It is not a unique solution of Permian stratigraphers to select a suitable marine sequence as the substitute of the traditional standard of non marine facies. The Downtonian Series of continental succession in the classic area was officially replaced by the Pridolian Series of marine facies as the fourth series of the Silurian System in the early 1980s. Similarly, the Upper Carboniferous is going to be defined in the marine sequences in palaeoequatorial areas rather than the continental sequences in classic areas.

Stability of Nomenclature
In his comment, Lazarev warns the unsuitability of names of the Permian and its constituent subdivisions. He expresses apprehension that the acceptance of the names Guadalupian and Lopingian and the logic of modernization of the scale will eventually lead refusal of the name "Permian" (28).

The instability of the system's name seems to be overstated. On the contrary, the name "Permian" will definitely be more stable as a consequence of establishment of the GSSPs for the systemic boundaries. No sign can be detected to change the name of the system because of using Guadalupian and Lopingian in the integrated scheme. The replacement of the Downtonian Series of Silurian classic succession with the Pridolian Series in the 1980s has not caused such a problem to the system's name at all.

The stability of stratigraphic names relies largely on the principle of priority. We agree with the statement in "The Stratigraphic Guide" (41) that priority of nomenclature should be respected, however, the critical factors should always be the usefulness of the unit, the adequacy of its description, freedom from ambiguity, and suitability for widespread application.

In the proposed scheme, the names of chronostratigraphic units of the traditional succession are adopted as far as possible so that the priority has been fully considered. For instance, the Kungurian sequence of the traditional standard is obviously not suitable for defining the GSSPs, but a compromise was made to reserve the name since the basal part can be approximately recognized in terms of a conodont zone.

Correlation

Lack of directly worldwide applicable biostratigraphic zonation of Late Permian implies that we will have tremendous work in ensuing from the establishment of GSSPs for intra-systemic boundaries.

We fully agree with the suggestion to integrate into the chronostratigraphic scheme with the data of magnetostratigraphy (33) and did incorporate the magnetostratigraphic zones with biostratigraphic sequences in the proposed scheme (17). Since the proposed scheme relies heavily on the biostratigraphic successions of palaeoequatorial areas, the radiometric age, the magnetostratigraphic and eventostratigraphic data will play a major role in erecting the North-South correlation of the Permian. The Subcommission needs to address this issue in its future planning. At present, emphasis of these studies should be put on locating the Illawarra Reversal and dating the volcanic beds in the succession of precise biostratigraphic control in palaeoequatorial areas. Improved correlation between zones of conodonts, fusulinids and ammonoids and their calibration to the magnetostratigraphic and radiochronological time scale will provide a framework to link all regions both inside and beyond the Palaeoequatorial realm.

SUBDIVISIONS AND BOUNDARY LEVELS

The Cisuralian Series

The name was proposed by Waterhouse (47) to denote an interval from the Asselian, the Sakmarian to the Artinskian. The name of the Uralian Series employed by Jin et al. (17) which has been used to denote various time interval, and the Yukian Series brought up again by Leven (27) are both not accepted.

Definitions of the constituent stages are suggested as follows based on the biostratigraphic successions of their eponymous areas (4). The initial Sakmarian Stage boundary lies at the base of the *Streptognathodus postfusus* in the Shihanskian Horizon. The initial Artinskian Stage boundary lies at the base of the *Sweetognathus whitei* Zone within the Bursevsk Horizon Formation. The Kungurian Stage was restricted formerly to the Philipovian and Irenian horizons of the type area. However, it has been proposed redefinition of the lower boundary at the base of the Sarginskian Horizon, originally included within the Artinskian Stage. This is a readily correlatable level, marked by the first appearances of the fusulinacean *Parafusulina*, the ammonoids *Propinacoceras* and

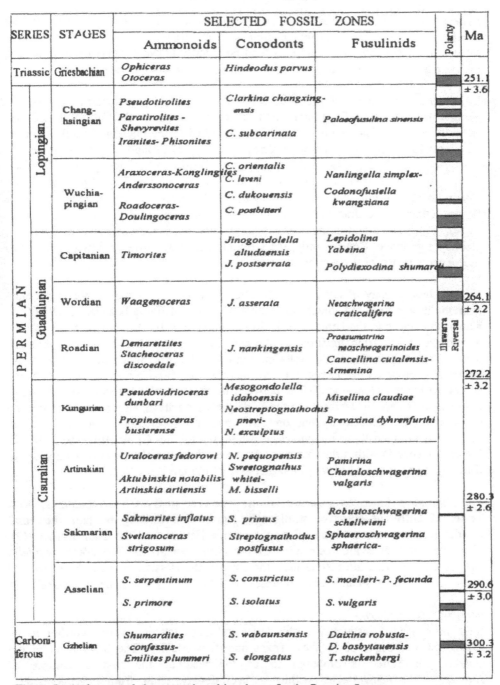

SERIES		STAGES	SELECTED FOSSIL ZONES: Ammonoids	Conodonts	Fusulinids	Polarity	Ma
Triassic		Griesbachian	*Ophiceras* *Otoceras*	*Hindeodus parvus*			251.1 ± 3.6
PERMIAN	Lopingian	Chang-hsingian	*Pseudotirolites* *Paratirolites - Shevyrevites* *Iranites- Phisonites*	*Clarkina changxing-ensis* *C. subcarinata*	*Palaeofusulina sinensis*		
PERMIAN	Lopingian	Wuchia-pingian	*Araxoceras-Konglingites* *Anderssonoceras* *Roadoceras-Doulingoceras*	*C. orientalis* *C. leveni* *C. dukouensis* *C. postbitteri*	*Nanlingella simplex-Codonofusiella kwangsiana*		
PERMIAN	Guadalupian	Capitanian	*Timorites*	*Jinogondolella altudaensis* *J. postserrata*	*Lepidolina* *Yabeina* *Polydiexodina shumardi*		
PERMIAN	Guadalupian	Wordian	*Waagenoceras*	*J. asserata*	*Neoschwagerina craticalifera*	Illawarra Reversal	264.1 ± 2.2
PERMIAN	Guadalupian	Roadian	*Demaretzites* *Stacheoceras discoedale*	*J. nankingensis*	*Praesumatrina neoschwagerinoides* *Cancellina cutalensis-Armenina*		272.2 ± 3.2
PERMIAN	Cisuralian	Kungurian	*Pseudovidrioceras dunbari* *Propinacoceras busterense*	*Mesogondolella idahoensis* *Neostreptognathodus pnevi-* *N. exculptus*	*Misellina claudiae* *Brevaxina dyhrenfurthi*		
PERMIAN	Cisuralian	Artinskian	*Uraloceras fedorowi* *Aktubinskia notabilis-* *Artinskia artiensis*	*N. pequopensis* *Sweetognathus whitei-* *M. bisselli*	*Pamirina* *Charaloschwagerina valgaris*		280.3 ± 2.6
PERMIAN	Cisuralian	Sakmarian	*Sakmarites inflatus* *Svetlanoceras strigosum*	*S. primus* *Streptognathodus postfusus*	*Robustoschwagerina schellwieni* *Sphaeroschwagerina sphaerica-*		
PERMIAN	Cisuralian	Asselian	*S. serpentinum* *S. primore*	*S. constrictus* *S. isolatus*	*S. moelleri- P. fecunda* *S. vulgaris*		290.6 ± 3.0
Carboni-ferous		Gzhelian	*Shumardites confessus-* *Emilites plummeri*	*S. wabaunsensis* *S. elongatus*	*Daixina robusta-* *D. bosbytauensis* *T. stuckenbergi*		300.3 ± 3.2

Figure 2. An integrated chronostratigraphic scheme for the Permian System.
The stratigraphic ranges of fossils zones reflect a general version of respective series and stages but are mainly determined based on zonation in their eponymous regions. However, the Kungurian and Guadalupian fossil zones combine the conodont zones and ammonoid assemblages of Southwestern USA with the fusulinid zones of South China.

Neocrimites, and the conodont *Neostreptognathodus pequopensis*. In the present scheme, the base of the *Neostreptognathodus pnevi* Zone of the Saraninsk Horizon is selected for definition, as it represents the first significant evolutionary event following the introduction of *N. pequopensis* which occurs at a major sequence boundary.

In the original version of proposed scheme (17), the corresponding unit of the Kungurian Stage was defined under the name of the Chihsian \ the Cathedralian Series since these two regional sequences are both qualified as the intercontinental standard for a post-Cisuralian Lower Permian series. The name of Leonardian was suggested for this series in the revised version of the scheme (19). Finally, the Permian Subcommission decided to retain the name of the Kungurian Stage from the classic Cisuralian standard instead of the Chihsian \ the Cathedralian Series or the Leonardian Series. However, the time span of this stage is much longer than the other Permian stages, and eustatic and biotic changes around the base of the Kungurian Stage are globally significant. Consequently, the Cisuralian may be further subdivided into two independent series, or two subseries.

Nevertheless, paucity of more useful biostratigraphic groups such as the conodonts, ammonoids and fusulinids precludes the potentiality of defining boundary levels of the Kungurian stages in the Urals. The Tethyan sequence of the Kungurian Stage is well documented by the fine zonation of the and the conodonts. It was divided into two regional stages referable to the Bolorian and the Kubergandinian Stage. The Bolorian Stage was proposed by Leven (26) to include the *Misellina* Genozone with the stratotype located along the divides between the Tschalidala, Kundara and Zidadara rivers in southwestern Darvase, and the Kubergandinian Stage, also by Leven to include the *Armenia* and *Cancellina* Genozone, with reference section along the right bank of the Kubergandinian River in southeastern Pamir. In South China, the base of the *N. insculptus* Zone is within the *Pamirina* Zone and therefore, the basal boundary of the Kungurian Series could be lower than that of the Bolorian Stage (43).

The alternative American sequence has its type area in the Glass Mountains, West Texas of USA, in objective stratigraphic succession directly beneath the basal Guadalupian. Component regional stages are the Hessian and the Cathedralian (40). Five conodont zones are recognized in the type Leonardian. The *Mesogondolella gujiaoensis - N. exculptus* Zone near the base of the Hessian contains *N. pnevi* and other conodonts that should permit precise correlation to *N. pnevi* Zone in the Urals.

The Guadalupian Series

It has been decided to standardize this series and its constituent stages according to the biostratigraphic sequence in West Texas and New Mexico. The basal level of the Guadalupian Series is proposed to be indicated by the first appearance of *Jinogondolella nankingensis* within a evolutionary cline from *Mesogondolella idahoensis* to *J. nankingensis* in the El Centro Member of the Cutoff Formation. So far as I understand, its corresponding zone of other fossil groups can not be directly fixed in its proposed type section. Moreover, this level has not yet been recognized in Tethyan successions. In South China, the lowest level of *J. nankingensis* present known is the *Praesumatrina neoschwagerinoides* Zone, but the Roadian ammonoid fauna is said to be found from the *Cancellina cutalensis* Zone (2). Proposed initial boundaries of the Wordian and the Capitanian Stage are respectively indicated by the first appearance of *J. aserrata* within the Gateway Limestone Member of the Brushy Canyon Formation and that of *J. postserrata* within the Pinery Member of the Bell Canyon Formation (11).

The Lopingian Series

Among the most popular names for the uppermost series such as the Lopingian Series (16), the Dzhulfian Series (9), the Transcaucasian Series, the Yichangian Series (47) and others, the Lopingian appears to be the first formally nominated series name for this part with reference to a relatively complete marine succession. The establishment of a complete succession of conodont zones from the Capitanian Stage to the Wuchiapingian Stage of the Lopingian Series enforced the international momentum to use the Lopingian Series and its constituent stages as the international standard for the last series of the Permian System (18, 30, 31, 32). The base of *C. postbitteri* zone represents a turning point from *Jinogondolella* to *Clarkina*, and so it is the most promising stratigraphic level for the Guadalupian \ Lopingian boundary. It is proposed to place this boundary within the top part of bed 19 in the Penglaitang Section, Laibin County of Guangxi. The fossil assemblages corresponding with the *C. postbitteri* Zone include the *Roadoceras-Doulingoceras* Zone (18,33).

Two stages, the Wuchiapingian and the Changhsingian stages are included in this series. Zhao et al. (48) formally proposed the D Section in Meishan of Changxing County as the stratotype of the Changhsingian Stage, while the base of the stage is located at the base of bed 2, the horizon between the *Clarkina orientalis* zone and the *Clarkina subcarinata* Zone. The basal part of this stage is also marked by the occurrence of the advanced forms of *Palaeofusiella*, and of the ammonoids of Tapashanitidae and Pseudotirolitidae. The Dzhulfian and Dorashamian Stage are corresponding respectively with the Wuchiapingian and Changhsingian Stage. However, the basal part of the Dzhulfian Stage and the top part of the Dorashamian Stage are not fully developed in their type area comparing to the standard succession in South China.

The Tatarian Stage of the traditional standard comprises the uppermost part of the Guadalupian since the Illawarra Reversal appears in the basal part of the Tatarian and the Capitanian (34). The Lopingian marine deposits in Pangea can be determined by occurrence of the *Cyclolobus* fauna as a reliable mark since it has been confirmed by associated Lopingian conodonts and foraminifers of Tethyan aspect in the Salt Range (46).

The conodonts from the Guadalupian-Lopingian boundary sequences used to be included in a single ill-defined zone, the *Clarkina bitteri-C.? welcoxi* Zone (23), or the *Neogondolella liangshanensis - N. bitteri* Zone. In Southwest USA, Kozur dated the conodonts, including *Mesogondolella altudaensis, M. "babcocki*, and *Clarkina subcarinata* from the Altuda Formation as the Dzhulfian-Lower Changhsingian (23). The establishment of a succession of 6 conodont zones unknown previously between the *postserrata* Zone of Capitanian and the *liangshanensis* Zone of middle Wuchiapingian in South China (30, 31 32) proves that the "Dzhulfian-Lower Changhsingian conodonts" of Kozur (23) should be late Capitanian and earliest Wuchiapingian in age. It also provides with a solid basis for defining the Guadalupian-Lopingian boundary. Despite of various arguments subsequent to the papers of Mei et al., this zonation of Guadalupian-Lopingian conodonts has been essentially accepted (24). However, Kozur (24) argued that "Clearly seen in western Texas, *M. prexuanhanensis* developed from *M. shannoi* and appeared considerably before *C. altudaensis*, whereas *"xuanhanensis"* (= *M. nuchalina*) occurs there within the lower *C. altudaensis* Zone". As indicated by Mei and Wardlaw (33), the occurrences of these species recorded in the explanation of plate 4 (24) are in contradiction with his statement. A succession from the *J. altudaensis*, the *J. prexuanhanensis* to the *J. xuanhanensis* Zone is clearly exhibited in his own records.

Because of the priority of the Dzhulfian Stage over the Wuchiapingian Stage, the attempt has been made to replace the Wuchiapingian Stage by two stages, namely the Laibinian Stage (18, 43) below and the Dzhulfian Stage above, the initial boundary of the latter is changed to the base of the *Cl. leveni* Zone. This suggestion was put forward under condition of that to define the stratotype of the Dzhulfian Stage in South China. The Permian Subcommission agreed with taking the advantages of integrating the type regions as minimum as possible, and therefore, declined to keep the name of the Dzhulfian Stage.

Magnetostratigraphic Sequence
A major part of Permian magnetostratigraphic sequence is assigned to the Carboniferous-Permian Reversed Megazone (CPRM), and the rest to the Permian-Triassic Mixed Megazone (PTMM). Menning (34) distinguished 5 normal zones for the Permian part of the CPRM.

The boundary between these two megazones, i.e. the Illawarra Reversal (IR) is located in the upper part of the Maokou Formation (Late Guadalupian) in South China (15), the Lower part of the Wargal Formation in Salt Range (13), which are corresponding with the *Neoschwagerina margaritae* Zone, the Lower Tatarian in the Urals (25), the Yats Formation in SW USA and therefore, is usually dated as Late Guadalupian in age. Recently, it was also defined in the Wittingham Coal Measure in the Northwestern Sydney Basin (45) and the Upper Shihotse Formation in North China (8). Based on these data, a possible age of the Ufimian was given. But, it can also be re-interpreted its age as the Late Guadalupian (21, 43).

Some 15 normal zones belonging to the Permian part of the PTMM were recognized. There are two and a possible other one normal zones in the late Guadalupian Series. The Wuchiapingian Stage contains two normal zones close its lower part, and one upper in Salt Range and South China. At the Meishan Section, the Changhsingian Stage comprises a normal zone at the base, at least three mixed zones in the lower part, a normal zone in the upper and a reversal near the top (29).

Isotopic Age

The isotopic age of Permian boundaries is now under an overall re-study on the volcanic intercalations with fine biostratigraphic control by several labs. The age for the Carboniferous-Permian boundary is estimated as 296 Ma by Menning (34) and as 298 Ma by Roberts et al. (39) based on the age 298.7 ± 8 Ma and 297.8 ± 8 Ma from the tuffs from the basal Rotliegend of Saar-Nahe Basin, Germany. An age of 300.3 ± 3.2 Ma for a tuff constrained by the fusulinid *Jigulites jigulensis* Zone in the middle of Gzhelian Stage in the Urals is of particular importance in estimating the age of Carboniferous-Permian boundary (5).

The age of the Asselian *Sphaeroschwagerina moelleri- Pseudofusulina fecunda* Zone is 290.6 ± 3.0 Ma, that of the *Pseudofusulina uralensis* Zone of the Uppermost Sakmarian is 280.3 ± 2.6 Ma and that of the *P. pedissequa* Zone of Lowermost Artinskian is 280.3 ± 2.4 Ma in the Urals (39).

Radiometric ages from Eastern Australia, 272.2 ± 3.2 Ma for the tuffs of Greta Coal Measures of Early Kungurian, 264.1 ± 2.2 Ma for the Mulbering Siltstone of Early Ufimian and 253.4 ± 3.2 Ma for the Ingelera Formation of Kazanian need to be re-interpreted. An alternative correlation between the Australian and the traditional Permian successions suggests to refer the Greta Coal Measures to the Early Ufimian, the Mulbering Siltstone to the Late Guadalupian and the Ingelera Formation to the Late Wuchiapingian(21). This suggestion is further proved by a reliable date of 264.1 ± 2.2 Ma for a bentonite bed just below the proposed base of the Capitanian stratotype and that of 252.0 ± 0.3 Ma for the base of the Changhsingian Stage in the Meshan Section, South China (S. Bowling, personal communication, 1996).

An age of 251.1 ± 3.6 Ma for the Permian - Triassic boundary clay bed of the Meishan Section, South China has been confirmed recently (38) as the latter gave a 249.9 ± 1.5 Ma for the same bed.

CONCLUSIONS

An integrated scheme is desirable since the Subcommission could not put into active cooperative projects on defining the various boundary stratotypes of a global standard time scale without such a primary scheme. Regardless of the divergence in some minor points of definitions of subdivisions, the proposed scheme as a whole is sufficient to permit the Subcommission to step in the stage of selecting the GSSPs for various intra-systemic boundaries. At least, the Subcommission can now focus its efforts on erecting the GSSPs for the basal boundary of the Guadalupian, the Lopingian as well as the Kungurian.

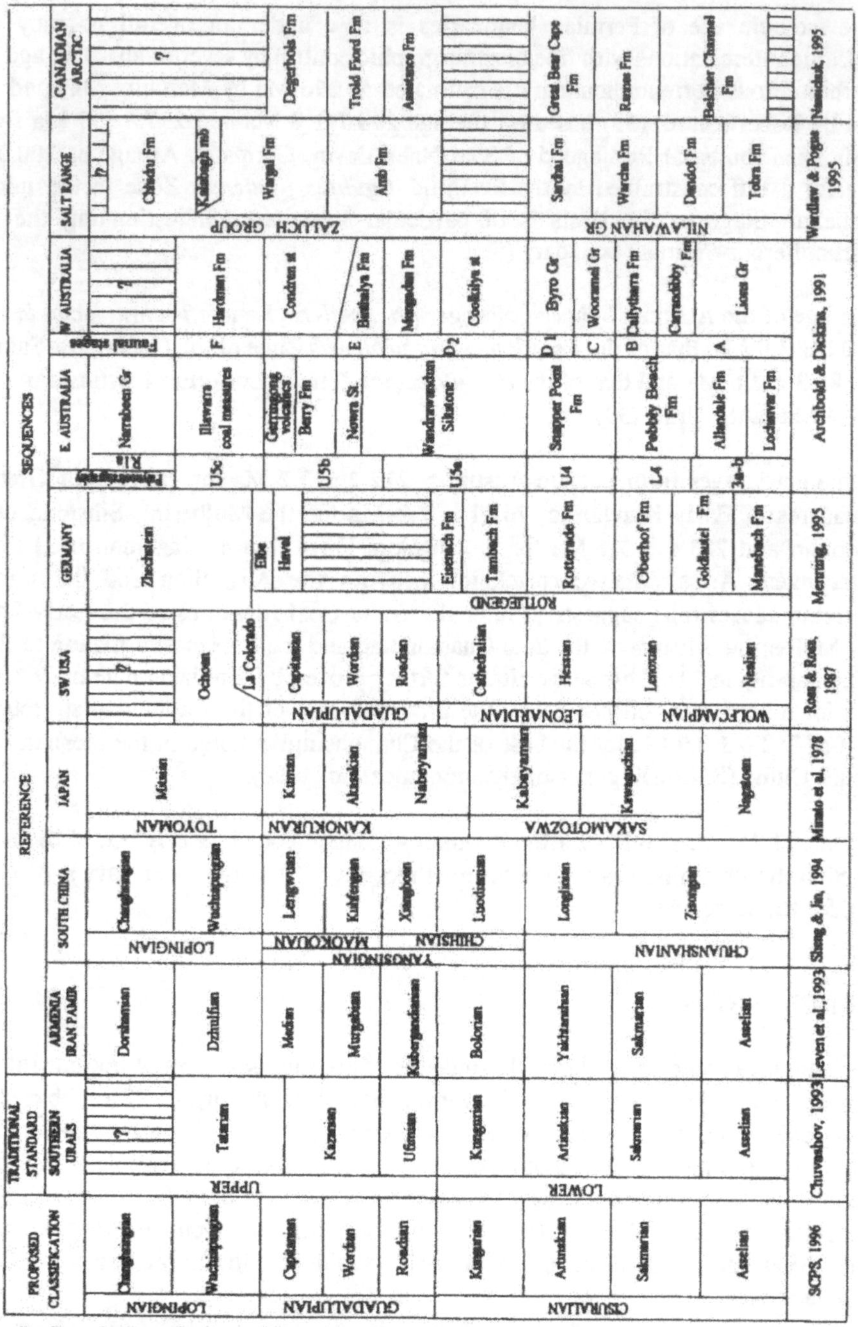

Figure 3. Correlation of selected Permian successions

The regional successions on the Text-figure 3 are adopted from the following authors' contributions: Germany from Menning (34); Southwestern USA from Ross and Ross (40); Western and Eastern Australia from Archbold and Dickins (1); the Urals from Chuvashov (4), the Salt Range from Wardlaw and Pogue (46), the Kitakami Mts. of Japan from Minato et al (35) and the Arctic from Nassichuk (36).

REFERENCE S

1. Archbold, N. W. and J. M. Dickins. *Australian Phanerozoic Timescales: 6. A standard for the Permian System* in *Australian. Bureau of Mineral Resources, Australian Record* 1989. 36 (1991).
2. Bogoslovskaya, M.F. and Leonova, T.B. Comments on the Proposed Operational Scheme of Permian Chronostratigraphy, *Permophiles*. 25, 15-17 (1994).
3. Burov Boris V. and Natalia K. Esaulova. On the Problems of the Study of the Upper Permian Stratotypes, *Permophiles*, 27, .30-34 (1995).
4. Chuvashov B. I. and Nairn, A. E. M.. Permian System: Guides to Geological Excursions in the Uralian Type localities. *Occasional publications ESRI*, new ser. 10, 1-303. University of South Carolina (1993).
5. Chuvashov, B.I., Foster, C.B., Mizens, G.A., Roberts, J., Claoue-Long, J.C. Radiometric (Shrimp)dates for some biostratigraphic horizons and event levels from the Russian and Eastern Australian Upper Carboniferous and Permian, *Permophiles*. 28, 29-36 (1996).
6. Davydov, V. I., B. F. Glenister, C. Spinosa, S. M. Ritter, V.V. Chernykh, B.R. Wardlaw, W. S. Snyder. Proposal of Aidaralash as GSSP for the base of the Permian System, *Permophiles*. 24, 1 - 9 (1994).
7. Davydov, V. I. On the Proposed Operational Scheme of Permian Chronostratigraphy, *Permophiles*. 25, 20-21 (1994).
8. Embleton, B. J. J., McElhinny, M.W., Ma, X.H., Zhang, Z.K. & Li, X.L. Permo-Triassic magnetostratigraphy in China: the type section near Taiyuan, Shanxi Province, North China, *Geophys. J. Int.* 126, 382-388 (1996).
9. Furnish, W. M. Permian Stage names. In Logan and Hills (ed.) The Permian and Triassic Systems and their mutual boundary, *Canadian Society of Petroleum Geologists, Memoir* 2, 522—549 (1973).
10. Glenister, B. F. and Furnish, W. M. The Permian Ammonoids of Australia, *Journal of Palaeontology*. 35, 673—736 (1961).
11. Glenister, B F., D. W. Boyd, W. M. Furnish, R. E. Grant, M. T. Harris, H. Kozur, L. L. Lambert, W. W. Nassichuk, N. D. Newell, L. C. Pray, C. Spinosa, B. R. Wardlaw, G. L. Wilde and T.E. Yancy. The Guadalupian: Proposed international standard for a Middle Permian Series, *International Geology Review*, 34:9, 857—888 (1992).
12. Grunt, Tatjana. Standard Permian Biostratigraphic Scale with Respect to Permian Marine Biogeography, *Permophiles*. 25, 21-25 (1994).
13. Haag, M. & Heller, F.. Late Permian to Early Triassic magnetostratigraphy, *Earth Planet. Sci. Letter*. 107, 42-54 (1991).
14. Harland, W. B., Armstrong, R. L., Cox, A. V., Craig, L. E., Smith, A. G., and Smith, D. G. *A Geological Time Scale 1989*, 1—263. Cambridge University Press (1990).
15. Heller, F., Chen, H. H., Dobson, J. & Haag, M. Permian-Triassic magnetostratigraphy — new results from South China, *Phys. Earth Planet. Int.*, 89, 281-295 (1995).
16. Huang T. K. The Permian formations of Southern China. *Memoirs of the Geological Survey of China, Ser.* A: 10, 1—140 (1932).
17. Jin Yu-gan, B. R. Glenister, C. K. Kotlyar & Sheng Jin-zhang. An operational scheme of Permian chronostratigraphy , *Palaeoworld*. .4, 1-14 (1994).
18. Jin Yu-gan,, Mei Shi-long & Zhu Zili. The Maokouan-Lopingian boundary sequences in South China, *Palaeoworld*. 4, 119-132 (1994).
19. Jin Yu-gan, Sheng Jin-zhang, Glenister, B.F., Furnish, W.M., Kotlyar, G.V., Wardlaw, B.R., Heinz Kozur, Ross, C.R. and Claude Spinosa. Revised Operational Scheme of Permian Chronostratigraphy, *Permophiles*. 25, 12-14 (1994).
20. Jin Yu-gan. A global chronostratigraphic scheme for the Permian System, *Permophiles*. 28, 4-10.

21. Jin Yu-gan and M. Menning. A possible   North-South Correlation of the Permian,
    Permophiles(in press).
22. Kotlyar Galina V. Comments on the Proposed Operational Scheme of Permian
    Chronostratigraphy, *Permophiles*. 26, 23-25.
23. Kozur, H. Dzhulfian and Early Changxingian (Late Permian) Tethyan conodonts from the
    Glass Mountains, West Texas, *N. JB. Geol. Palaeont. Abh.*, 187:1, 1-114 (1992).
24. Kozur, H. Permian conodont zonation and its importance for the Permian stratigraphic
    standard scale, *Geol. Palant. Mitt. Innsbruck*. 20, 165-205 (1995).
25. Khramov, A.N. Palaeomagnetic investigations of Upper Permian and Lower Triassic sections
    on the northern and eastern Russian Platform. *Trudy VNIGRI, Nedra, Leningrad*, 204, 145-
    174 (1967).
26. Leven, E. Y. *Explanatory Notes to the Permian stratigraphic scale of the Tethyan Realm*,
    VSEGEL, Leningrad (1980).
27. Leven, E Y. Comments on the Proposed Operational Scheme of Permian Chronostratigraphy.
    Permophiles. 25, 17-20 (1994).
28. Lazarev, S.S. The Problem of Stability of Stratigraphic Names in the Standard of the Permian
    System. *Abstract, International Symposium on Evolution of Permian Marine Biota, Moscow*.
    60-61 (1995).
29 .Li Hua-mei and Wang Jun-da. Magnetostratigraphy of Permo-Triassic boundary section of
    Meishan of Changxing, Zhejiang. *Science in China* B: 6, 652—658 (1989).
30. Mei Shi-long, Jin Yu-gan and Bruce R. Wardlaw. Succession of conodont zones from the
    Permian "Kuhfeng Formation", Xuanhan. Sichuan and its implications in global correlation.
    Acta Palaeontologica Sinica, 42:1, 1—21(1994).
31 .Mei Shi-long, Jin Yu-gan and Bruce R. Wardlaw. Succession of conodont zones from
    Permian Wuchiaping Formation, Northern Sichuan, *Acta Micropalaeontologica Sinica*. 2, 1
    —21 (1994).
32 Mei Shi-long, Jin Yu-gan and Bruce R. Wardlaw. Zonation of conodonts from the Maokouan-
    Lopingian boundary strata South China, *Palaeoworld*. 4, 165-178 (1994).
33 .Mei Si-long and Bruce R. Wardlaw. On the Permian conodont "Liangshanensis-bitteri" Zone
    and related problems. in Wang and Wang (Eds.), *Centennial Memorial Volume of Prof. Sun
    Yun-zhu*, 130-140, China Univ. of Geosciences Press, Wuhan, China (1996).
34. Menning, M. A Numerical Time Scale for the Permian and Triassic Periods. An integrated
    Time Analysis. in Scholle, Peryt, T.M. & Ulmer-scholle (Eds.), *Permian of the Northern
    Pangea*. 1, 77-97, Springer-Verlag, Berlin (1995).
35. Minato, M. Kato, M., Nakamura, K., Hasegawa, Y., Choi, D. R. and Tazawa, J., 1978.
    Biostratigraphy and correlation of the Permian in Japan, *Journ. Fac. Sci., Hokkaido Univ.*,
    Ser. 4, 18:1-2, 11—47 (1978).
36. Nassichuk, W. W. Permian Ammonoids in the Arctic Regions of the world, in Scholle, Peryt,
    T.M. & Ulmer-scholle (Eds.), *Permian of the Northern Pangea*. 1, 210-236 Springer-Verlag,
    Berlin (1995).
37. Remane, J., M.G. Bassett, J. W. Cowie, K. H. Gohrbandt, H. R. Lane, O. Mechelsen and
    Wang Naiwen. Guidelines for the establishment of global chronostratigraphic standard by the
    International Commission on Stratigraphy (ICS), *Episodes*.19:3, 77-81 (1997).
38. Renne, P. R., Zhang Zichao, M. A. Richard, M. T. Black & A. R. Basu. Synchrony and causal
    relations between Permian -Triassic boundary crises and Siberian flood volcanism, *Sciences*.
    269, 1413-1416 (1995).
39. Roberts, J. Claoue'-Long, J. C. Foster C.B. SHRMP zircon dating of the Permian System of
    eastern Australia. *Australian J. Earth Sci.*. 43, 401-421 (1996).
40. Ross, C. A. and Ross, J. R. P. Late Paleozoic sea levels and depositional sequences, *in* Ross,
    C. A., and Haman, D., (Eds.), Timing and depositional history of eustatic sequences,

Constraints on seismic stratigraphy. *Cushman Foundation for Foraminiferal Research, Special Publication,* **24**, 137—149 (1987).

41. Salvador, A. (Ed.) *International Stratigraphic Guide, a guide to stratigraphic classification, terminology, and procedure.* 2nd edition, Int'l Union Geol. Sci. & Geol. Soc. Amer., Inc. 1-214 (1995)

42. Shevelev, A.I., Khalimbadzha, V.G., Burov, B.V., Esaulova, N.K. and Gusev, A.K. On the Problems of the International Permian Stratigraphic Scale, *Permophiles.* **27**, 38-39 (1994).

43. Sheng Jin-zhang and Jin Yu-gan. Correlation of Permian deposits of China, *Palaeoworld.* **4**, 14 -94 (1994).

44. Stepanov, D. L. The Permian System in the U. S. S. R. in Logan and Hills (Eds.) *The Permian and Triassic Systems and their mutual boundary, Canadian Society of Petroleum Geologists, Memoir* **2**, 120—137 (1973).

45. Theveniaut, H., Klootwiijik, C., Foster, C. & Giddings, J. Magnetostratigraphy of the Late Permian coal measures, Sydney & Gunnedah Basin: A regional & global correlation tool. *Proc. 28th Newcastle Symposium on "Advance in the study of the Sydney Basin", 15-17 April, 1994, Univ., Newcastle, Newcastle, N.S.W.* 11-23 (1994).

46. Wardlaw, B. R. and K. R. Pogue. The Permian of Pakistan. in Scholle, Peryt, T.M. & Ulmer-scholle (Eds), *Permian of the Northern Pangea,* **2**, 215-225. Springer-Verlag, Berlin (1995).

47. Waterhouse, J. B. An early Djulfian (Permian) brachiopod faunule from Upper Shyok Valley, Karakorum range, and the implications for dating of allied faunas from Iran and Pakistan, *Contribution to Himalayas Geolog.* **2**, 188—233 (1982).

48. Zhao, J. K., Sheng, J. Z. , Yao, Z. Q., Liang, X. L. , Chen, C. Z. , Rui, L. and Liao, A. T. The Changhsingian and Permian -Triassic boundary of South China, *Bulletin of the Nanjing Institute of Geology and Palaeontology, Academia Sinica.* **2**, 1—112 (1981).

Proc. 30ᵗʰ Int'l. Geol. Congr., Vol. 11 , pp. 115-119
Wang Naiwen and J. Remane (Eds)
© VSP 1997

High Frequence Glacio-Eustasy and Carbon Isotope Evolution of the *Triticites* Zone in South China

LIU BEN-PEI
Department of Geology. China University of Geosciences, 29 Xueyuan Road, Beijing 100083, P.R.China

LI RU-FENG
Department of Geosciences, University of Petroleum, Changping, Beijing 102200, P.R.China

Abstract

The detailed analysis of sequence stratigraphy and the systematical measurement of carbon isotope($\delta^{13}C$) have been done in the *Triticites* Zone (Kasimovian - Gzhelian) of the typical Carboniferous section exposed in the Guizhou province in South China. Two sequences and seventeen parasequences, which can be correlated with the two sequences and seventeen subsequences in the North American Midcontinent, have been distinguished in the *Triticites* Zone. This gives us a convincing evidence of the global synchroneity of the depositional records. The same high frequence glacio-eustacy is main causes of the above facts. Furthmore, the internal relations between $\delta^{13}C$ and eustasy have been studied, and the evolutionary regularities of $\delta^{13}C$ in a depositional sequence have been summarized. The research indicates that the $\delta^{13}C$ evolution in the carbonate sequence stratigraphy was controlled mainly by the periodic variation of global glacio-eustasy.

Keywords: sequence stratigraphy, carbon isotope evolution, glacio-eustasy, Triticites Zone, South China

INTRODUCTION

It is well known that Late Carboniferous was a stage of icehouse effect. Due to the weakening of intraplate rifting, the Yangtze Plate, which is the main part of the South China, entered a stable tectonic stage in Late Carboniferous. This research was carried out in the Dushan area of Guizhou locating on the Yangtze Plate in South China (Fig. 1), the eponymous area of Carboniferous System in China.

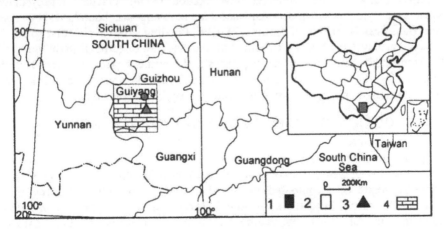

Figure 1. Map showing the location of the study area. 1. Location of the study area; 2. Study area; 3. Dushan; 4. Carbonate platform

Upon internationally accepted standard, the Kasimovian - Gzhelian stages accommodate the *Triticites* zone (*sensu lato*), underlain by *Fusulinella-Fusulina* zone and overlain by *Pseudoschwagerina-Sphaeroschwagerina* zone. In the study area, the strata of Kasimovian - Gzhelian age refer to the *Triticites* zone, dating from 295 to 290 Ma [4] .

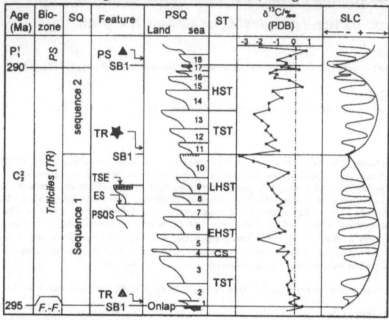

Figure 2. Sequence stratigraphy and δ¹³C evolution in the *Triticites* zone of the study area. SQ: sequence; PSQ: parasequence; ST: systems tract; SLC: sea-level curve; TR: *Triticites;*PS: *Pseudoschwagerina;* F.-F.: *Fusulinella - Fusulina;* PSQS: parasequence surface; ES: exposed surface; TSE: transgressive surface of erosion. Triangles indicate TR (PS) appearing; Star indicates TR thriving; CS: condensed section. Positive and negtive sign indicate rising and falling of sea-level respectively.

Characterized by monotonous carbonate association in the study area, sequence stratigraphy analysis starts with the investigation of microfacies and cyclicity of carbonates in order to recognize sequence and parasequence boundaries by exposed surface, transgressive surface of erosion (TSE)[9], flooding surfaces (fs), and so on. Then the depositional systems tracts are defined on the stacking patterns of parasequence sets [6,7]. Accordingly, two third order sequences and seventeen parasequences have been recognized (Fig. 2). This article is directed to the depositional sequence, glacio-eustasy, and the variation of carbon isotope composition in the *Triticitices* zone .

HIGH FREQUENCE GLACIO-EUSTASY IN *TRITICITES* ZONE

Microfacies and Features of Parasequences
The cyclic parasequences of the *Triticites* zone are well developed in the study area . A complete parasequence is composed, in ascending order, of biodetritus-bearing micrites, solitary coral-bearing biosparites, fusulinid-bearing biosparites and algal-laminated micrites from subtidal zone of open carbonate platform to tidal flat and supratidal zone.

All of the parasequences possess the characteristics of punctuated aggradational cycle (PAC) [3], which represents an upward-shallowing succession, beginning with rapid deepening at the base marked by a flooding surface and terminated by supratidal exposure (Fig. 3). It is worthy noting that the sea-level oscillation of rapid rising and slow falling exhibited by the PAC in the *Triticites* zone agrees well with the patterns of rapid thawing and slow growth of terristrial glaciation.

Depositional Systems Tracts and Sequence Boundary

The stacking pattern of parasequences and their relationship with depositional systems tracts in the two third order sequences of the *Trticities* zone demonstrate that : (1) Parasequences in the transgressive systems tracts (TST) increase in thickness upward and are characterized by higher ratio of subtidal micrites to tidal flat algal-laminated micrites, indicating a retrogradation pattern and increase of the accomodation space upward (parasequence 1-3 and 11-13); (2) Highstand systems tracts (HST) are well developed on carbonate platform , and can be divided into early highstand systems tracts (EHST) and later highstand systems tracts (LHST) based on the stacking pattern of parasequence sets . EHST are dominated by thick subtidal micrites (parasequence 5-6), LHST by lower ratio of subtidal micrites to tidal limestones (parasequence 7-10) ; (3) Each sequence merely includes TST and HST on the inner carbonte platform .

The two sequences in the *Triticities* zone belong to type I sequence, and the lower boundary of the sequence 1 is indicated by an onlapping surface created by the falling and succeeding rising of glacio-eustasy driven by Gondwana glacial events. The upper boundary of the sequence 1 is at the top of parasequence 10, marked by algal-laminated micrites with bird-eye structures and representing a supratidal exposure, associated with a distinct negative anomaly of $\delta^{13}C$. The lowest part of the sequence 2 consists of rudstones, which is the lag deposit in subtidal zone during rapid transgression. The sequence 2 is capped by the incomplete parasequence 17, resulting from the combination of regressive and succeeding transgressive erosions (Fig. 2).

Correlation and Mechanism of Depositional Sequence

Comparison of the depositional sequences in the *Triticites* zone in South China with that in North American Midcontinent demonstrates that: (1) Two third order sequences have been developed in the two areas [2,7]; (2) Seventeen fourth or fifth order transgressive-regressive (T-R) cycles have been recorded in the two areas [7,8] , and all of them are PACs of rapid transgressive and slow regressive characteristics and lack the regressive deposits of the toppest T-R cycle [1,7]. Such a comparison confirms that the two sequence and seventeen T-R cycles in the *Triticites* zone are traceable both on the South China and on North American Midcontinent.

It is well known that the frame of continents in the Late Carboniferous was of the embryonic form of the Pangea. The similar paleolatitude position of the two plates mentioned above connotes that they had experienced the same influence of the glacio-eustasy related to the waxing and waning of Gondwana ice sheets, so they have the same number of depositional sequences. Nevertheless, the difference in tectonics and paleogeography between South China and North American Midcontinent has resulted in the distinction of the seventeen T-R cycles in content .The North American Midcontinent was situated near the foreland of the Appalachia-Wichita Mountains [5]. The subsidence of higher magnitude accompanied with sufficient terrigenous influx led to the formation of subsequence in a fourth or fifth order sea-level change period. The Dushan area in South China had a stable tectonic background, where a monotonous petrographic association of carbonate platform far from terrigenous dilution was developed. A fourth or fifth order sea-level oscillation merely produced a parasequence.

CARBON ISOTOPE EVOLUTION IN THE DEPOSITIONAL SEQUENCE

Fifty-eight carbonte samples were systematically collected from the *Triticites* zone of the study area. The results of serial determination of $\delta^{13}C$ are shown in Figure 2.

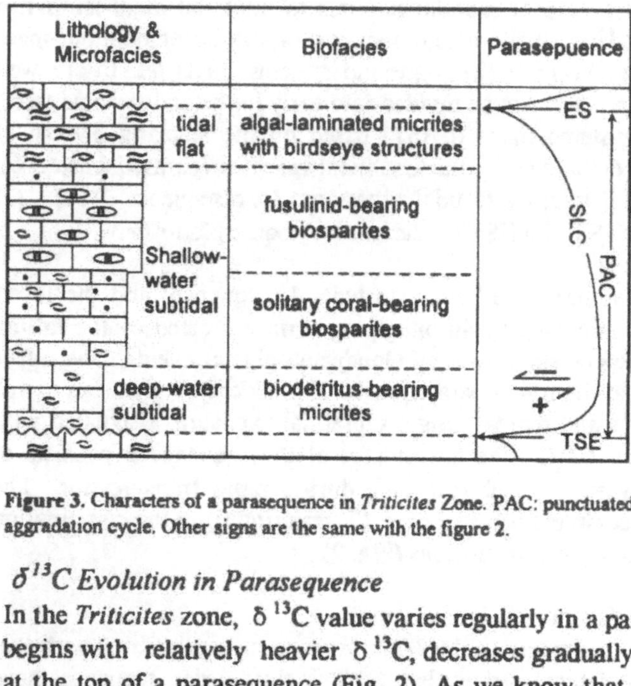

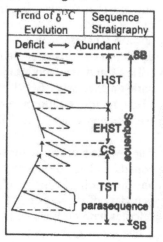

Figure 3. Characters of a parasequence in *Triticites* Zone. PAC: punctuated aggradation cycle. Other signs are the same with the figure 2.

Figure 4. Model of $\delta^{13}C$ evolution in the carbonate sequence stratigraphy

$\delta^{13}C$ Evolution in Parasequence

In the *Triticites* zone, $\delta^{13}C$ value varies regularly in a parasequence. Each parasequence begins with relatively heavier $\delta^{13}C$, decreases gradually upward and reaches minimum at the top of a parasequence (Fig. 2). As we know that the ^{13}C is primarily related to redox condition and the organic matter content of depositional environment. It is mentioned above that each parasequence experienced a process of rapid rising to gradual falling of sea-level change. During the sea-level rising, a relative deep environment provdied with an optimum condition for organic matter deposition, buring and preservation, and prevented them from oxidation, so that a high $\delta^{13}C$ value is thus recorded. With sea-level falling, decreased buring rate of organic matter and frequent exposure oxidation resulted in the light $\delta^{13}C$ value.

$\delta^{13}C$ Evolution in Depositional Systems Tract

A line connecting each intersection point of the trend line of $\delta^{13}C$ values in parasequence with the parasequence boundaries is drawn to delineate the evolutional trend of $\delta^{13}C$ values in depositional systems tracts. In TST, the slope of the line is less than 90° , indicating that the $\delta^{13}C$ has a general trend to be enriched upward. In HST, the slope of the line is larger than 90° , indicating that the $\delta^{13}C$ tends to be depleted upward (Fig.4). These results are identical with the sea-level change during transgressive and highstand period .

$\delta^{13}C$ Signatures of Sequence Boundary

The tops of the two sequences are made up of parasequence 10 and 17 respectively. The former consists of algal-laminated micrites with bird-eye structures of supratidal environment. The latter is characterized by relict wearthering bauxitic crust of 0.1m in thickness. Both of the sequence boundaries are marked by negative anomaly of $\delta^{13}C$, the former is distinct, but the later is indistinct because transgressive erosion had partly truncated the top of parasequence 17 (Fig.2).

CONCLUSIONS

Two sequences and seventeen parasequences, which can be correlated with the two sequences and seventeen subsequences in the North American Midcontinent, have been distinguished in the *Triticites* Zone of typical Carboniferous section exposed in Guizhou of South China[2,8]. The study shows that a third order sequence or even a high frequency T - R cycle can be traceable globally if it is formed in a stable tectonic background influenced mainly by glacio-eustasy, but its content is still subject to the interplay of multi-factors. We hold strongly that the depositional records of Kasimovian - Gzhelian ages are of global synchronism. $\delta^{13}C$ evolutional trend in marine carbonates does reflect the sea-level changes and facilitates the recognition of sequence (or parasequence) boundary, therefore, it is an effective method in the analysis of carbonate sequence stratigraphy.

ACKNOWLEDGMENTS

The research is conducted as a part of the research program of sequence stratigrphy supported by the State Commission of Science and Technology(P.R.C.) etc. We thank Prof. Shi Xiaoying and Doctor Jin Zhengkui for their careful review of the manuscript.

REFERENCES

1. Boardman II , D. R. et al.. Glacial-eustacy of the Upper Carboniferous-Lower Permian boundary strata in the North American Midcontinent. In:Beauchamp B, Embry A, eds. *Carboniferous to Jurassic Pangea, Program & Abstracts.* Canadian Society of Petroleum Geologists, 1993. 28 (1993).
2. Busch, R. M. and Rollins, H. B. Correlation of Carboniferous strata using a hierarchy of transgressive-regressive units, *Geology*, v.12, p.471-474 (1984).
3. Goodwin, P. W. and Anderson, E. J. Punctuated aggradational cycles: a general hypothesis of episodic stratigraphic accumulation, *Journal of Geology*, v. 93, p.515-553 (1985).
4. Harland, W. B. et al. *A geologic time scale.* London, Cambridge University Press, p. 19-23 (1982).
5. Heckel, P. H. Sea-level curve for Pennsylvanian eustatic marine transgressive-regressive depositional cycles along Midcontinent outcrop belt, North America, *Geology*, v. 14, p. 330-334 (1986).
6. Li Rufeng and Liu Benpei. Application of Carbon and Oxygen isotopes to carbonate sequence stratigraphy: an analysis of the Maping Formation, Southern Guizhou, China, *Earth Science--Journal of China University of Geosciences*, v.21(3). p. 261-266 (1996).
7. Liu Benpei and Li Rufeng. Carboniferous sequence stratigraphy and glacio-eustasy of *Triticites* Zone in Southern Guizhou, China, *Earth Science--Journal of China University of Geosciences*, v.19(5), p.553-564 (1994).
8. Ross, C. A. and Ross, J. R. *Late Paleozoic sea level and depositional sequences.* Cushman Foundation for Foraminiferal Research, Special Publication 24, p.137 (1987).
9. Weimer, R. J. Relation of unformities, and sea level changes, Cretaceous of Westen Interior, U. S. A. In: J. S. Schlee, ed. *Interregional uniformities and hydrocarbon accoumulation.* AAPG Memoir 36, p. 7-35 (1984).

Proc. 30ᵗʰ Int'l. Geol. Congr., Vol. 11, pp. 121-131
Wang Naiwen and J. Remane (Eds)
© VSP 1997

Devonian-Carboniferous Boundary in the Neritic Facies Areas of South China from the Viewpoint of Integrative Stratigraphy

WANG XUNLIAN   ZHANG SHIHONG   XUE XIAOFENG

Department of Geology, China University of Geosciences, Beijing, 100083, China

Abstract

In regard to isochroneity of stratigraphical unit boundaries, the present stratigraphical methods may be divided into two types. One is biostratigraphy, which is based on the irreversible evolution of life and is basic in stratigraphy, although its unit boundaries are in most cases diachronous. The other includes event stratigraphy, sequence stratigraphy, ecostratigraphy, magnetostratigraphy, seismic stratigraphy, and stable isotope stratigraphy, in which the unit boundaries are by definition isochronous and correlatable over a considerable extent, even all over the world. However, these units lack unique signals and are recurrent in geological history. Their age can not be decided without biostratigraphy or isotope dating. Integrative stratigraphy attempts to make full use of stratigraphic data, including physical, chemical and biological, and specially emphasizes a combined use of all stratigraphical methods. Within the stratigraphic framework established by means of biostratigraphy, event stratigraphic and other isochronous stratigraphic unit boundaries are used for more detailed division and precise correlation. It provides an effective method in high-resolution stratigraphy. In the present paper an attempt is made to discuss the Devonian - Carboniferous Boundary (DCB) in the neritic facies areas of South China from the viewpoint of integrative stratigraphy.

A combined study of biostratigraphy, event stratigraphy, sequence stratigraphy, and ecostratigraphy shows that in the neritic facies areas of South China, the DCB, matching the boundary between *Siphonodella praesulcata* Zone and *S. sulcata* Zone in pelagic facies areas, is not only higher than the top of the *Cystophrentis* Range Zone, but also higher than the top of the event beds which caused the extinction of *Cystophrentis*. We may here define the Devonian-Carboniferous Boundary by the most distinct transgressive surface within the *Cystophrentis\Pseudouralina* Interval Zone in the neritic facies areas of South China. This boundary is the base of transgress systems tract of a Sloss sequence, which coincides with the top surface of the DCB event beds, and approximately corresponds to the bottom of *Pseudouralinia* Assemblage Zone. In cyclic stratigraphy, this boundary is also the basal boundary of a T-R cycle. Remarkable changes in palaeoecologic pattern and sudden variation in magnetic susceptibility of limestones and stable isotope took place at the boundary, which seems therefore to be a ideal natural boundary between Devonian and Carboniferous.

Keywords: integrative stratigraphy, high-resolution stratigraphy, Devonian and Carboniferous, boundary, South China

DEFICIENCIES OF BIOSTRATIGRAPHY AND THE RAPID DEVELOPMENT OF INTEGRATIVE STRATIGRAPHY

Establishment of regional and global high-resolution chronostratigraphic framework and more precise correlation are fundamental work in stratigraphy. Up to present, almost all GSSP of Phanerozoic are determined mainly based on biostratigraphy. Biostratigraphy seems to have reached its limits of effective resolution, but a large number of problems with regard to correlation remain unsolved. Indeed, biostratigraphy has played an important part in stratigraphic correlation, but has its limits for the following deficiencies. First, it is very difficult, ever impossible, to determine the base boundary of a biozone, i.e. the first appearance datum of a zone-fossil, merely on the basis of biostratigraphy. Second, biostratigraphic unit boundaries are in most cases diachronous. When biozone unit boundaries are used as isochronous surface, the error may reach nearly a whole biozone. Probably they are isochronous only in case of regional or global mass extinction. Third, as no living beings could adapt to all kinds of environments, exact correlation between different facies areas is almost impossible. Fourth, in the course of preservation, biozones may contain mixed or merged fossils, which may cause wrong stratigraphic correlation. Long term divergence in stratigraphic correlation results primarily from the fact that diachronous biostratigraphic unit boundaries are used as isochronous indicator in stratigraphic correlation.

Integrative stratigraphy attempts to make full use of stratigraphic data, including physical, chemical and biological, and specially emphasizes combined study by using all stratigraphical methods. In addition to biostratigraphy, lithostratigraphy, and chronostratigraphy, other methods, including event stratigraphy, sequence stratigraphy, ecostratigraphy, stable isotope stratigraphy, seismic stratigraphy, and magnetostratigraphy, have been employed more and more with success in practice. All these methods are complementary and could be tested and verified with each other, thus providing an effective tool for precise correlation.

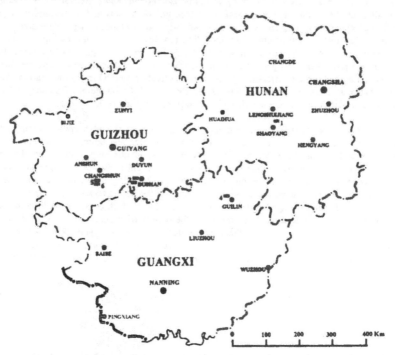

Figure 1. Map showing the location of sections. 1. Malanbian (central Hunan); 2. Baihupo (Dushan, southern Guizhou); 3. Qilinzhai (Dushan, southern Guizhou); 4. Nanbiancun (Guilin, Guangxi); 5. Dapoushang (Changshun, Guizhou); 6. Muhua (Changshun, Guizhou)

In regard to isochroneity of stratigraphical unit boundaries, the present stratigraphical methods may be divided into two types. One includes event stratigraphy, sequence stratigraphy, ecostratigraphy, magnetostratigraphy, seismic stratigraphy, and stable isotope stratigraphy, in which the unit boundaries are by definition   isochronous and correlatable over wide extent. However, these units lack unique signals and are recurrent in geological history. Their age can not be decided without biostratigraphy or isotope geochronology. The other is   biostratigraphy based on the principle of irreversible evolution of life, and therefore not recurrent in geological history. Thus biostratigraphy is basic in stratigraphy, and pay a key role in establishing stratigraphical framework, although its unit boundaries are in most cases diachronous. In general, only within a stratigraphical framework established on biostratigraphy, can other methods be validly applied to more precise division and correlation.

Continuous deposits of Upper Devonian and Lower Carboniferous are widespread in South China. In pelagic facies areas, the Nanbiancun Section of Guilin, and the Muhua Section and the Daposhang Section of Guizhou Province (Fig. 1), the complete lineage from *S. praesulcata* to *S. sulcata* was found, and the base of the Carboniferous is marked by the first appearance of the conodont species *Siphonodella sulcata* in the lineage from *S. praesulcata* to *S. sulcata*. However, different opinions exist on the DCB in the neritic facies areas of central Hunan, southern Guizhou and other areas, which is put at: 1) the base of *Cystophrentis* Range Zone, i.e. the top of the Shaodong Formation; 2) the top of *Cystophrentis* Rang Zone, marked by the extinction of *Cystophrentis*; 3) the base of the Shaodong Formation; and 4) the base of *Pseudouralina* Assemblage Zone. The divergence results from the two facts: 1) the strata of neritic facies could not be directly correlated with that of pelagic facies and all the stratotypes on DCB were chosen in pelagic facies areas; 2) the boundaries marked by different fossils are inconsistent owing to the varied rates of their evolution. It is evident that the stratigraphic correlation between neritic and pelagic facies seems to be very difficult merely on the basis of biostratigraphy. In the present paper an attempt is made to discuss the DCB in neritic facies areas of South China from the viewpoint of integrative stratigraphy.

BIOSTRATIGRAPHY ACROSS THE DCB IN SOUTH CHINA

Conodonts

C_1	Upper	*Siphonodella duplicata* Zone
	Lower	*Siphonodella duplicata* Zone
		Siphonodella sulcata Zone
D_3	Upper	*Siphonodella praesulcata* Zone
	Middle	*Siphonodella praesulcata* Zone
	Lower	*Siphonodella praesulcata* Zone

The conodont zones are mainly developed in pelagic facies areas, including the Nanbiancun Section, Guilin[3] (Fig. 2), the Muhua Section[1] and the Dapoushang Section, Guizhou[2], occasionally with some of the elements in shallow water facies area represented by the Xiakou Section, Guangxi[9], where the conodonts may be associated with foraminifera.

Rugose corals

C_1	*Pseudouralina* Assemblage Zone
D_3	*Cystophrentis\Pseudouralina* Interval Zone
	Cystophrentis Range Zone
	Caninia dorlodoti Assemblage Zone
	Ceriphyllum elegontum Assemblage Zone

The rugose coral zones occur only in neritic facies areas such as southern and central Hunan, northern Guangdong, northern Guangxi and southern Guizhou (Figs. 1, 2). They can hardly be correlated directly with the conodont zones in pelagic facies.

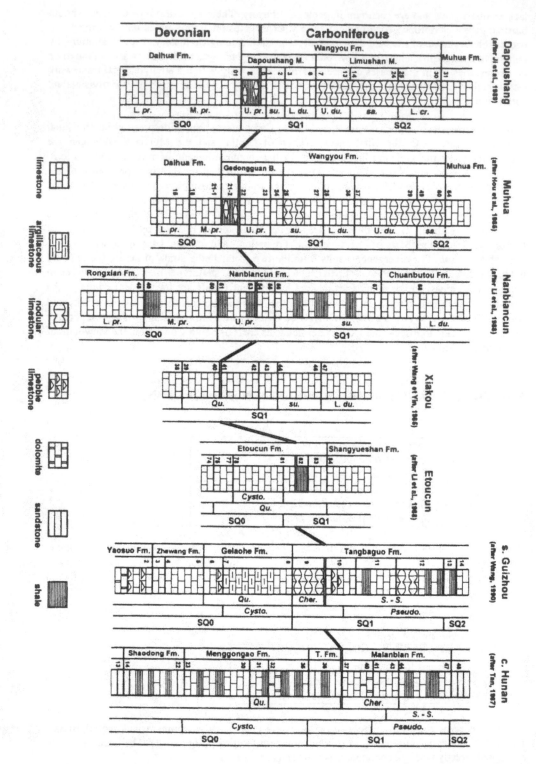

Figure 2. The Devonian-Carboniferous Boundary in South China. L. *pr.*, Lower *praesulcata* Zone; M. *pr.*, Middle *praesulcata* Zone; U. *pr.*, Upper *praesulcata* Zone; *su.*, *sulcata* Zone; L. *du.*, Lower *duplicata* Zone; U. *du.*, Upper *duplicata* Zone; *sa.*, *sandbergi* Zone; L. *cr.*, Lower *crenulata* Zone; *Qu.*, *Quasiendothyra konensis-Qu. kobeitusana* Assemblage Zone; *Cher.*, *Chernyshinella* Assemblage Zone; *S.- S.*, *Septabrunsiina krainica-Septatournayella segmentata* Assemblage Zone; *Cysto.*, *Cystophrentis* Range Zone; *Pseudo.*, *Pseudouralina* Assemblage Zone. SQ0, Sequence 0; SQ1, Sequence 1; SQ2, Sequence 2. T. Fm., Tianxinxiang Formation.

Foraminifera

C_1	*Septabrunsiina krainica-Septatournayella segmentata* Assemblage Zone
D_3	*Chernyshinella* Assemblage Zone
	Quasiendothyra konensis-Qu. kobeitusana Assemblage Zone
	Septatournayella rauserae-Quasiendothyra communia Assemblage Zone
	Umbellina gravis-U. spinosa Assemblage Zone

The genus *Quasiendothyra*, with high evolution rate, is widespread and has played an very important role in setting the DCB. Recently, the extinction of *Quasiendothyra* (the top of the *Qu. konensis-Qu. kobeitusana* Assemblage Zone) is regarded as an indicator of DCB in neritic facies areas. This horizon is lower than the base of the *S. sulcata* Zone in the Xiakou Section[9] (Fig. 2), but higher than the top of the *Cystophrentis* Range Zone in the Etoucun Section of Guilin[4]. In southern Guizhou[10], southern Hunan[7] the horizon is higher than top of the *Cystophrentis* Range Zone but lower than the base of the *Pseudouralina* Assemblage Zone (Fig. 2). Evidently, according to such a correlation, the DCB may better be drawn between *Cystophrentis* Range Zone and *Pseudouralina* Assemblage Zone, i.e. the *Cystophrentis\Pseudouralina* Interval Zone, in the neritic facies areas of South China. More precise correlation seems to be difficult merely on the basis of biostratigraphy.

DCB GEOLOGICAL EVENT AND ITS EVIDENCE IN SOUTH CHINA

As early as in 1976, Wu Xianghe suggested that there was an important depositional break between the *Cystophrentis* Range Zone and *Pseudouralina* Assemblage Zone in southern Guizhou, with distinct changes in sedimentology and palaeontology, and took the break as the natural DCB. In 1983 Sandberg pointed out that there was a short pulse of the eustatic fall event which made the sea shallower and caused sedimentary discontinuity near the DCB. He referred this as an event boundary. This event has been confirmed in North America[5, 13] and the France-Belgium Basin[6].

In recent years, the eustatic fall event near DCB has been well documented in South China. The event beds occur at the top of middle *praesulcata* Zone in the Muhua Section (Bed 22-1)[1] (Fig. 2), and at the base of the upper *praesulcata* Zone in both the Dapoushang Section (Bed E)[2] and the Nanbiancun Section (Beds 51-53)[3]. In other areas of the world, the horizon of the DCB event beds also range from the top of the middle *praesulcata* Zone to the base of the upper *praesulcata* Zone. In a word, all the event beds are lower than the base of the *S. sulcata* Zone in pelagic facies areas, and occur within *S. praesulcata* Zone.

The eustatic fall event has also been recorded in the *Cystophrentis\Pseudouralina* Interval Zone in the neritic facies of South China, which resulted in the extinction of rugose coral *Cystophrentis*, brachiopods *Cyrtospirifer* and *Tenticospirifer*. In southern Guizhou the event bed lies at the base of the Tangbagou Formation (Bed 9)[12]. In central Hunan, it is characterized by the Tianxinxiang Formation (Bed 36)[7] consisting of sandstone and shale intercalated between limestones.

Chronostratigraphy		Sequence	Littoral and platform facies			Slope facies	Intraplatform basin facies	
			c. Hunan	s. Guizhou	Rugose Coral Zone	Guilin	Muhua	Dapoushan
Carboniferous	Tournaisian	SQ3	Shidengzi Fm. / Doulingao Fm. (LST / TST)	Xiangbai Fm.	*Kryssettingeophyllum* Subzone	Longkou Fm.	Dawuba Fm. / Muhua Fm.	Muhua Fm.
		SQ2	Tianciping Fm. (SMST / LST / TST)	Tangbagou Fm.	*Pseudouralinia Assemblage Zone*	Chuanbutou Fm.		
		SQ1	Malanbian Fm. (SMST / LST / TST)		*Pseudouralinia tangpakouensis* Subzone		Wangyou Fm.	Wangyou Fm.
Devonian	Strunian	SQ0	Tianzhukang Fm. (SMST / LST) Menggongao Fm. / Gelaohe Fm. (TST) Zhewang Fm. (TST) Shaodong Fm. (LST) Oujiachong Fm. Yaosuo Fm. Xikuangshan Fm. (LST / SB1 / LST)	Zhewang Fm.	*Cystophrentis* / *Pseudouralinia* Interval Zone; *Cystophrentis* Range Zone; *Caninia dorlodoti* Assemblage Zone; *Ceriphyllum elegantum* Assemblage Zone	Nanbiancun Fm. / Rongxian Fm. / Dalhua Fm.	Dalhua Fm.	Dalhua Fm.

Conodont zones (Guilin / Muhua / Dapoushan):

- (*Scaliognathodus anchoralis* Zone)
- (*Gnathodus typicus* Zone)
- (Upper *crenulata* Zone)
- (Lower *crenulata* Zone)
- *sandbergi* Zone
- Upper *duplicata* Zone
- Lower *duplicata* Zone
- *sulcata* Zone
- Upper *praesulcata* Zone
- Middle *praesulcata* Zone
- Lower *praesulcata* Zone (*expansa* Zone)

Table 1. Correlation of the Upper Devonian and Lower Carboniferous sequence stratigraphy, lithostratigraphy, chronostratigraphy and biostratigraphy

To sum up, the horizon of the sea level fall event near the DCB is lower than the boundary between *S. praesulcata* and *S. sulcata* zones, but higher than the top of the *Cystophrentis* Range Zone. The boundary between *S. praesulcata* and *S. sulcata* zones is at least higher than the top of the *Cystophrentis* Range Zone.

SEQUENCE STRATIGRAPHY ACROSS THE DCB IN SOUTH CHINA

Theoretically, sequence boundary and systems tract bounding surface have chronostratigraphic significance. Therefore the depositional sequences may offer a good means for the stratigraphical correlation in different facies areas.

Four depositional sequences were recognized in the upper part of the Upper Devonian and Tournaisian of South China [11]. They are named SQ0, SQ1, SQ2 and SQ3 in ascending order. SQ0 is Strunian (uppermost Devonian), and the other three Tournaisian in age (Table 1).

Study of sequence stratigraphy shows that in both pelagic facies and neritic facies areas the basal boundary of SQ1 is consistent with the base of the sea level fall event bed near DCB, and the event bed represents the shelf margin systems tract (SMST) of SQ1 [11]. Moreover, in slope and basin facies areas, the top of the event bed, i. e. top of the SMST in SQ1, is slightly lower than the base of *S. sulcata* Zone. In the neritic facies areas of South China, the DCB, matching the boundary between *S. praesulcata* Zone and *S. sulcata* Zone in pelagic facies areas, is not only higher than the top of *Cystophrentis* Range Zone, but also higher than the top of DCB event beds.

ECOSTRATIGRAPHY ACROSS THE DCB IN SOUTH CHINA

Ecostratigraphy provides an effective method for precise correlation of different facies areas. The eustatic fall event near DCB not only affected the ecosystem of neritic facies areas, but also disturbed the fossil succession of pelagic facies areas. With the eustatic fall, paleocommunities and conodont biofacies moved toward sea (Fig. 3). In central Hunan, the *Crurithyris – Quasiendothyra* Community of the upper part of the Menggongao Formation, consisting mainly of brachiopods, rugose corals, foraminifera, and other marine fossils, was replaced by the *Verrucosisporites-Vallatisporites-Retispora* Community containing only spores and broken plants in the Tianxinxiang Formation. In southern Guizhou, *Cystophrentis-Yanguania-Quasiendothyra* Community composed of abundant rugose corals, brachiopods, and foraminifera in the upper part of the Gelaohe Formation gave way to a foraminifera *Vicinesphaera-Bisphaera* Community of the base of the Tangbagou Formation[10]. In the latter the foraminifera are small in size, simple in structure, with long geologic range, usually occurring only in environment unfavorable to other marine fossils. The sea level fall event caused the extinction of *Cystophrentis*, *Cyrtospirifer*, and *Tenticospirifer* in widespread neritic facies areas, and relative shallow water conodont biofacies also commonly occurred in the fossil succession of pelagic facies areas. In pelagic facies areas, including the Nanbiancun Section of Guilin, the Muhua Section and the Dapoushang Section in Guizhou, shallow water facies *Protognathodus* Biofacies replaced the deep water facies *Palmatolepis* Biofacies (Fig. 3)[1, 2, 8]. After the fall event, with the sea level rise, the paleocommunities and conodont biofacies moved toward land, and new communities and conodont biofacies developed in various facies areas. The spore *Verrucosisporites-Vallatisporites-Retispora* Community gave way to *Vicinesphaera-Bisphaera-Syringopora* Community comprising marine fossils in central Hunan, while the foraminifera *Vicinesphaera-Bisphaera* Community of southern Guizhou was replaced by a low wave calm water environment *Retichonetes-Ptychomaloechia* Community. In the Nanbiancun Section of Guilin, the Muhua Section and the Dapoushang Section of Guizhou, deep water *Siphonodella-Polygnathus* Biofacies took the place of *Protognathodus* Biofacies. The remarkable replacement of communities and biofacies occur within the Upper *praesulcata* Zone in both the Nanbiancun Section and the Dapoushang Section, and is consistent with the base of the Upper *praesulcata* Zone in the Muhua Section. Therefore, the Devonian-Carboniferous Boundary in the neritic facies areas of South China is at least higher than the top of the event bed leading to the extinction of *Cystophrentis*.

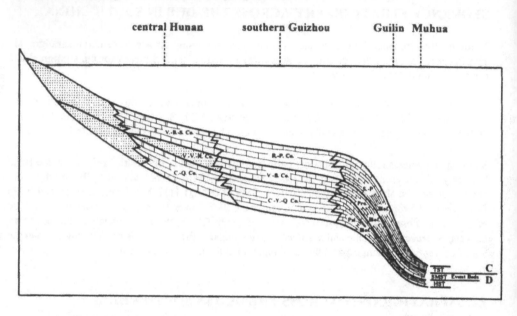

Figure 3. Sketch map showing distribution of the communities and biofacies near the Devonian-Carboniferous Boundary of South China and their relation to sedimentary sequences and regressive event. V.-B.-S. Co. *Vicinesphaera-Bisphaera-Syringopora* Community; V.-V.-R. Co. *Verrucosisporites-Vallatisporites-Retispora* Community; C.-Q. Co. *Crurithyris – Quasiendothyra* Community; R.-P. *Retichonetes-Ptychomaloechia* Community; V.-B. Co. *Vicinesphaera-Bisphaera* Community; C.-Y.-Q. Co. *Cystophrentis-Yanguania- Quasiendothyra* Community; S.-P. Biof. *Siphonodella-Polygnathus* Biofacies; Pro. Biof. *Protognathodus* Biofacies; Pal. Biof. *Palmatolepis* Biofacies.

VARIATION OF MAGNETIC SUSCEPTIBILITY OF LIMESTONES AND δ^{13}C NEAR THE DCB IN SOUTH CHINA

The remarkable variation in magnetic susceptibility and δ^{13}C of limestones took place near the DCB in South China. In the DCB event beds of southern Guizhou, the susceptibility of limestones are relatively lower and reach its lower limit, after then, the value of susceptibility of limestones increased suddenly (Fig. 4A). Same phenomenon is also observed in the Muhua Section. δ^{13}C of the event beds in southern Guizhou are relatively higher, above the top of the event bed, the value of δ^{13} C decreased significantly (Fig. 4B).

CONCLUSION AND DISCUSSION

A combined study of biostratigraphy, event Stratigraphy, sequence stratigraphy, and ecostratigraphy shows that in the neritic facies areas of South China the DCB, matching the boundary between *S. praesulcata* Zone and *S. sulcata* Zone in pelagic facies, is not only higher than top of the *Cystophrentis* Range Zone, but also higher than top of the event bed which results in the extinction of *Cystophrentis*. The physical boundary surface which is most close to the DCB determined by the first appearance of *S. sulcata* is the transgressive surface on the top of the sea level fall event bed. We may here define the Devonian - Carboniferous Boundary by the most distinct transgressive surface within the *Cystophrentis\Pseudouralina* Interval Zone in the neritic facies areas of South China. This boundary is base of transgress systems tract of a Sloss sequence, which coincides with the top surface of the DCB event beds, and approximately corresponds to the

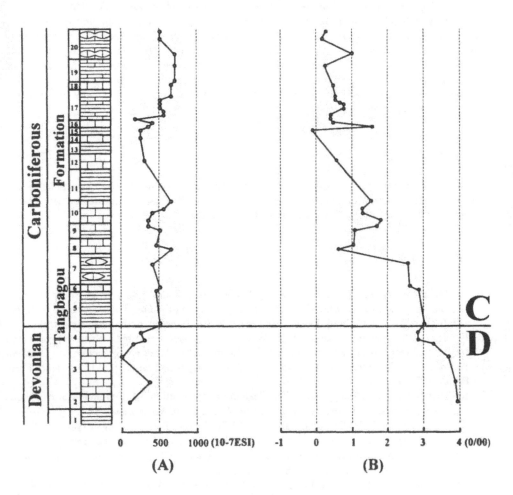

Figure 4. Magnetic susceptibility (A) and δ¹³ C (B) of limestones near the Devonian-Carboniferous Boundary of Dushan County, southern Guizhou

bottom of *Pseudouralinia* Assemblage Zone. In cyclic stratigraphy, this boundary is also the basal boundary of a T-R cycle. Important changes in palaeoecologic pattern and sudden variation in magnetic susceptibility and stable isotope of limestones took place at the boundary. This boundary seems to be a ideal natural boundary between Devonian and Carboniferous, applicable and easily recognized both in neritic and pelagic facies areas. It lies at top of the nodular limestones at the base of the Tangbagou Formation in South Guizhou, at the base of the Malanbian Formation in central Hunan. In Xiakou Section [9], it is drawn between the Bed 40 and the Bed 41, where sparite is suddenly replaced by micritic limestone, and in Etoucun section [4], between the Bed 81 and the Bed 82.

It has to be pointed out that the DCB suggested in this paper is not consistent with the biozones. In the pelagic facies areas, as mentioned above, the boundary lies in Upper *praesulcata* Zone or matches the base of Upper *praesulcata* Zone. In the neritic facies areas the boundary is close to the top of *Quasiendothyra konensis-Qu. kobeitusana* Assemblage Zone. In southern Guizhou and central Hunan the boundary is slightly higher, while in Xiakou and Etoucun, slightly lower than the top of *Quasiendothyra konensis-Qu. kobeitusana* Assemblage Zone. Considering that the distribution of fossils is controlled by sedimentary facies and biostratigraphic unit is diachronic, maybe the boundary on the ground of integrative stratigraphy is more reasonable than that merely based on biostratigraphy. Even in pelagic facies strata containing the conodont lineage from *S. praesulcata* to *S. sulcata*, the boundary proposed in this paper can still be used as the indicator of the DCB. Determining stratigraphical boundary based on integrative stratigraphy makes it possible to correlate the strata of different facies over a considerable distance, and greatly improves the precision of stratigraphical correlation.

ACKNOWLEDGMENT

The research is sponsored by the "SSLC", a key project supported by the State Science and Technology Commission of China and the Ministry of Geology and Mineral Resources. Thanks are given to Professor Wang Hongzhen for reading of the manuscript.

REFERENCES

1. Hou Hongfei, Ji Qiang, Wu Xianghe, Xiong Jianfei, Wang Shitao, Gao Lianda, Sheng Huaibin, Wei Jiayong, and Susan Turner. *Muhua section of Devonian-Carboniferous boundary beds.* pp. 1-226. Geological Publishing House, Beijing (1985). (in Chinese with English summary)
2. Ji Qiang, Wei Jiayong, Wang Zengji, Wang Shitao, Sheng Huaibin, Wang Hongdi, Hou Jingpeng, Xiang Liwen, Feng Rulin, and Fu Guomin. *The Dapoushang Section, An excellent section for the Devonian-Carboniferous boundary stratotype in China.* pp. 1-165. Science Press, Beijing (1989).
3. Li Zhengliang, Lu Hongjin, and Yu Changmin. Description of Devonian-Carboniferous boundary sections. In: *Devonian-Carboniferous Boundary in Nanbiancun, Guilin, China — — Aspects and Records.* Yu Changmin (Ed.). pp. 19-36. Science Press, Beijing (1988)
4. Li Zhengliang, Guo Shuying, He Guiying, and Yu Changmin. Regional stratigraphy. In: *Devonian-Carboniferous Boundary in Nanbiancun, Guilin, China — — Aspects and Records.* Yu Changmin (Ed.). pp. 9-18. Science Press, Beijing (1988).
5. C.A. Sandberg, R.C. Gutschick, J.G. Johnson, F.G. Poole, and W.J. Sando. Middle Devonian to Late Mississipian event stratigraphy of overthrust belt region, western United States. *Annales de la Societe Geologique de Belgique* 109, 205-207 (1986).
6. M.V. Steenwinkel. The Devonian-Carboniferous Boundary in the Vicinity of Dinant, Belgium. *Cour. Forsch.-Inst. Senckenberg* 67, 57-69 (1984).
7. Tan Zhengxiu. Stratigraphy. In: *The Late Devonian and Early Carboniferous strata and palaeobiocoenosis of Hunan.* Regional Geological Surveying Party, Bureau of Geology and Mineral Resources of Hunan Province (Ed.). pp. 2-65. Geological Publishing House, Beijing (1987). (in Chinese with English summary)
8. Wang Chengyuan and Yin Baoan. Conodonts. In: *Devonian-Carboniferous Boundary in Nanbiancun, Guilin, China — — Aspects and Records.* Yu Changmin (Ed.). pp. 105-148. Science Press, Beijing (1988).

9. Wang Chengyuan and Yin Baoan. A important Devonian-Carboniferous boundary stratotype in neritic faces of Yishan County, Guangxi. *Acta Micropalaeontologica Sinica* **2:1**, 28-48 (1985). (in Chinese with English summary)

10. Wang Keliang. On the Devonian-Carboniferous boundary based on foraminiferal fauna from South China. *Acta Micropalaeontology Sinica* **4:2**, 161-173 (1987). (in Chinese with English summary)

11. Wang Xunlian, Li Shilong, and Wang Yue. Upper Devonian and Lower Carboniferous sequence stratigraphy of South China. *Journal of China University of Geosciences* **7:1**, 87-94 (1996).

12. Wang Zengji. *The Carboniferous System of China.* pp. 1-419. Geological Publishing House, Beijing (1990). (in Chinese)

13. W. Ziegler and C.A. Sandberg. Important candidate sections for the stratotype of conodont based Devonian — Carboniferous boundary. *Cour. Forsch.-Inst. Senckenberg* **67**, 231-239 (1984).

Proc. 30ʰ Int'l. Geol. Congr., Vol. 11 , pp. 133-141
Wang Naiwen and J. Remane (Eds)

Dicynodon and Late Permian Pangea

SPENCER G. LUCAS

New Mexico Museum of Natural History, 1801 Mountain Road N. W., Albuquerque, New Mexico 87104, U.S.A.

Abstract

Dicynodon is a Late Permian therapsid reptile whose fossils occur in South Africa, Tanzania, Zambia, Scotland, Russia, China and Laos. Functional analysis indicates that *Dicynodon* was a terrestrial herbivore that lacked any aquatic capabilities. The distribution of *Dicynodon* is most easily understood by dispersal during the Late Permian across dry land routes. This implies land connections of the North China and Indochina microplates to the Pangean supercontinent.

Keywords: Dicynodon, Pangea, Late Permian, North China, Indochina

INTRODUCTION

Fossil vertebrates provide strong evidence that the continents drifted during geological time, so they have long been used to constrain plate tectonic reconstructions based on a variety of geological and geophysical data. Perhaps the best known case is that of *Lystrosaurus*, an Early Triassic dicynodont reptile whose fossils are known from South Africa, India, Antarctica, Russia and China. This distribution on currently widely separated continents is most easily explained by continental drift and the assembly of Early Triassic Pangea [9].

Dicynodon (Fig. 1) is a Late Permian dicynodont whose fossils have a broad geographic distribution that rivals that of *Lystrosaurus* (Fig. 2). This distribution of a terrestrial reptile is best explained by an integrated Late Permian Pangea. The distribution of *Dicynodon* thus challenges geological and geophysical data that suggest the North China and Indochina microplates were separated by oceanic barriers from Pangea during the Late Permian.

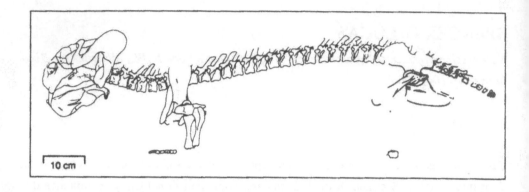

Figure 1. Left lateral view of the skeleton of *Dicynodon trigonocephalus* [20]. Note the absence of any evidence of aquatic capabilities by this fully terrestrial, quadrupedal herbivore.

MORPHOLOGY, PALEOBIOLOGY AND TAXONOMY OF *DICYNODON*

Dicynodon was a common and widespread Late Permian dicynodont (see below) of medium to large size (skull lengths range from 100 mm to more than 400 mm). Key features of *Dicynodon* (Fig. 1) include a single pair of maxillary tusks in the upper jaw, an edentulous lower jaw, a low boss formed by the nasals above the external nares, a ventral extension of the palatal rim to form the caniniform process, a sharp-edged continuous palatal rim with a notch, fused dentaries with narrow dentary tables followed by a deep dentary sulcus, a weak coronoid process, and a large mandibular fenestra bounded dorsally by a lateral dentary shelf. Postcranially, *Dicynodon* has a short, stout neck, a long, relatively inflexible back and a short tail. The limbs and limb girdles are robust, and the manus and pes are short and thick and have digits that bore hoof-like tips [7, 12, 20-21].

Functional analysis [12, 16, 20, 22] indicates that *Dicynodon* was a slow-moving, quadrupedal herbivore that undertook powerful mastication of tough and bulky plant foods. It was fully terrestrial and capable of sustained locomotory effort. No features of its skeleton suggest any arboreal or aquatic habits.

The specimen of *Dicynodon* described by King [20] has a skull about 200 mm long and a total length (tip of snout to tip of tail) of about 1 m. Scaling other *Dicynodon*, which are mostly known from isolated skulls, to this skeleton indicates that *Dicynodon* ranged in length from 0.5 m to more than 2 m. Hotton [16] estimated that the body mass of *Dicynodon* ranged from that of a beaver to a brown bear, or about 13 to 200 kg.

Haughton and Brink [15] listed 111(!) named species of *Dicynodon*. Keyser [19] and Cluver and Hotton [7] reduced this number, especially by reallocating some species to other Late Permian dicynodont genera such as *Oudenodon* and *Diictodon*. King [21] listed 59 species of *Dicynodon*, but noted that many may not be valid. I cannot resolve the species-level taxonomy of *Dicynodon*, but instead use the genus as the operational taxonomic unit of biogeographic analysis.

DISTRIBUTION OF *DICYNODON*

The distribution of *Dicynodon* (Fig. 2) establishes a *Dicynodon* biochron of Late Permian age [24] recognized at the following locations:

1. Karoo basin, South Africa, where specimens of *Dicynodon* (= *Daptocephalus*) are the dominant tetrapod fossils in the Upper Permian *Daptocephalus* Zone of Kitching [23] in the Barberskrans Member of the Balfour Formation (Beaufort Group).

2. The "lower bone bed" at Kingori in the Ruhuhu Basin of Tanzania [14]. Haughton [14] and Huene [17] named three species of *Dicynodon* (*D. huenei*, *D. tealei* and *D. bathyrhynchus*) from the Ruhuhu Basin.

3. The type of *D. roberti* [5] is from "Horizon 5" of Boonstra [6] in the Luangwa Valley 3-4 mi (4.8-6.4 km) north of Nt'awere. Zambia The

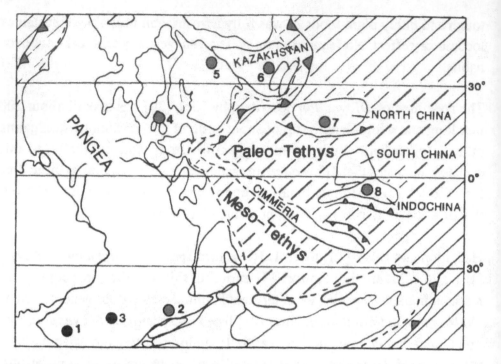

Figure 2. Reconstruction of Late Permian Pangea (26) showing *Dicynodon* occurrences:
1=South Africa, 2=Tanzania, 3=Zambia, 4=Scotland, 5=Russia, 6=Junggur Basin, China;
7=Ordos Basin, China, 8=Laos.

skeleton of *D. trigonocephalus* described by King [20] is from the
Madumabisa Mudstones of the Luangwa Valley [18].

4. Cutties Hillock Quarry, Elgin, Scotland [21, 27] is in the Cutties Hillock
Sandstone Formation [4] and has produced the specialized pareiasaur *Elginia
mirabilis* [27] and a supposed procolophonid [34] that may actually be
another pareiasaur [3]. Newton [27] named the Cutties Hillock Quarry
dicynodonts *Gordonia* and *Geikia*. King [21] synonymized *Gordonia* with
Dicynodon and recognized *D. traquairi* as the only valid species from Cutties
Hillock Quarry. Rowe [30] argued that *Geikia* is a distinct genus to which he
also assigned *"Dicynodon" locusticeps* [17] from Tanzania. Rowe
recognized a close relationship of *Geikia* to *Pelanomodon*, but King [21]
argued that *Geikia* is a synonym of *Pelanomodon*. There is consencus that

the Cutties Hillock Quarry tetrapod assemblage is of latest Permian age [3-4, 30, 34].

5. Northern Dvina fauna, near Kotlass, Russia, Tatarian Zone IV of Efremov [11]. Amalitzky [1] named *Gordonia annae* and other dicynodonts subsequently reassigned to *Dicynodon annae* by Sushkin [33]. Another Upper Tatarian occurrence of *Dicynodon* in Russia is a skull from the Vyatskyan horizon near Voskresensk on the Vetluga River, currently under study.

6. The Quanzijie, Wutonggou and Guodikeng Formations in the Junggur and Turpan Basins, Xinjiang Province, China. These formations have produced specimens assigned to *Dicynodon* as well as to its generic synonyms *Kunpania, Jimusaria, Striodon* and *Turfanodon* [21, 24, 31-32, 35]. Furthermore, a tuskless skull of a dicynodont named *Dicynodon tienshanensis* belongs to *Diictodon,* a genus also known from the uppermost Permian of South Africa [7].

7. The Naobaogou Formation at Shiguai, Nei Monggol, China, produced a skull named *Daqingshanodon limbus* [36]. Zhu's [36] diagnosis of this taxon mentions many features diagnostic of *Dicynodon*. It does not differentiate *Daqingshanodon* as a distinct genus, so I synonymize it with *Dicynodon*. The Naobaogou Formation occurrence of *Dicynodon limbus* thus extends the distribution of *Dicynodon* into the Ordos basin of the North China microplate.

8. From north of the Mekong River in the Luang-Prabang area of Laos, Battail et al. [2] reported the discovery of several skulls of *Dicynodon* in Counillon's [10] "fourth zone" of conglomerates, mudstones and sandstones ("zone des argiles violettes"). This locality supposedly produced the Early Triassic dicynodont *Lystrosaurus* [21, 28-29], but as Battail et al. [2] argued, the so-called *Lystrosaurus* specimen (now lost) probably also pertained to *Dicynodon*. This establishes the presence of *Dicynodon* on the Indochina microplate.

PALEOGEOGRAPHIC IMPLICATIONS

The occurrence of *Dicynodon* from these now dispersed localities forms a powerful argument for the assembly of Pangea by Late Permian time. Yet, the most recent plate-tectonic reconstructions of Late Permian Pangea show a clear marine separation of the North China and Indochina microplates from the rest of the supercontinent [e.g.,13, 25-26, 37] (Fig. 2). Four possible explanations of the distribution of *Dicynodon* can be formulated: vicariance, island hopping, swimming and dry land dispersal.

Vicariance

The distribution of *Dicynodon* is a generalized track that could be taken to indicate the genus originated before the Late Permian at a time when Pangea (or Gondwanaland) was assembled. A cladistic hypothesis of dicynodont phylogeny [8] indicates that *Dicynodon* shares its closest common ancestry with Triassic dicynodonts. The ancestor of this common ancestry was shared with the Late Permian genus *Aulacephalodon*. This means *Dicynodon* originated during the Late Permian. Furthermore, North China and Indochina were only part of a supercontinent (Gondwanaland) during the Cambro-Ordovician (Tremadoc); after that time they are considered to have been isolated microplates [26]. Vicariance of an earlier, pre-Late Permian distribution of *Dicynodon* thus cannot be used to explain its Late Permian distribution.

Island hopping

Dicynodon could have reached the apparently isolated North China and Indochina microplates by island hopping from the integrated Pangean landmass. Such a sweepstakes route of dispersal seems most plausible between Kazakhstan and North China, which were separated by only a short oceanic distance that probably contained an intervening island arc (Fig. 2). During the Late Permian, Indochina was near the eastern end of the Cimmerian continent, which had its western end near the Pangean nucleus (Fig. 2). If archipelagic conditions existed in either Paleo- or Meso-Tethys this would have facilitated island hopping.

Swimming

Island hopping might involve some swimming, or a long swim to North China and/or Indochina might explain the distribution of *Dicynodon*. However, functional analysis of *Dicynodon* suggests no aquatic capabilities. Indeed, the animal seems a most improbable swimmer.

Dry land dispersal

The distribution of *Dicynodon* is most easily understood by terrestrial dispersal during the Late Permian across dry land routes. This implies land connections of the North China and Indochina microplates to the Pangean supercontinent. It certainly is most parsimonious to argue that fully terrestrial *Dicynodon* of the Late Permian dispersed over a single landmass, as has been well argued for its successor, *Lystrosaurus* of the Early Triassic.

Acknowledgments

I thank I. Metcalfe for information and the National Geographic Society (Grant 5459-95) for partial support of this research.

REFERENCES

1. V. Amalitzky. Diagnoses of the new forms of vertebrates and plants from the Upper Permian of North Dvina, *Bull. Acad. Sci. St. Petersburg* **16**: 329-340 (1922).

2. B. Battail, J. Dejax, P. Richir, P. Taquet and M. Ve'ran. New data on the continental Upper Permian in the area of Luang-Prabang, Laos, *Geol. Surv. Vietnam J. Geol. B* **5-6**: 11-15 (1995).

3. M. J. Benton and P. S. Spencer. *Fossil reptiles of Great Britain*. Chapman & Hall, London (1995).

4. M. J. Benton and A. D. Walker. Palaeoecology, taphonomy and dating of Permo-Triassic reptiles from Elgin, north-east Scotland, *Palaeont.* **28**: 207-234 (1985).

5. L. D. Boonstra. A report on some Karoo reptiles from the Luangwa Valley, northern Rhodesia, *Quart. J. Geol. Soc. London* **94**: 371-384 (1938).

6. L. D. Boonstra. A report on a collection of fossil reptilian bones from Tanganyika Territory, *Ann. South Afr. Mus.* **42**: 5-18 (1953) .

7. M. A. Cluver and N. Hotton III. The genera *Dicynodon* and *Diictodon* and their bearing on the classification of the Dicynodontia, *Ann. S. Afr. Mus.* 83: 99-146 (1981).

8. M. A. Cluver and G. M. King. A reassessment of the relationships of Permian Dicynodontia (Reptilia, Therapsida) and a new classification of dicynodonts, *Ann. S. Afr. Mus.* 83: 99-146 (1983).

9. E. H. Colbert. Tetrapods and continents, *Q. Rev. Biol.* 46: 250-268 (1971).

10. H. Counillon. Documents pour servir a l'étude géologique des environs de Luang-Prabang (Cochinchine), *Comp. Rend. Acad. Sci. Paris* 123: 1330-1333 (1896).

11. I. A. Efremov. Preliminary description of new Permian and Triassic terrestrial vertebrates from the USSR, *Trudy. Paleon. Inst.* 10: 1-140 (1940).

12. R. F. Ewer. Anatomy of the anomodont *Daptocephalus leoniceps*, *Proc. Zool. Soc. London* 136: 375-402 (1961).

13. J. Golonka, M. I. Ross and C. R. Scotese. Phanerozoic paleogeographic and paleoclimatic modeling maps, *Canad. Soc. Petrol. Geol. Mem.* 17: 1-47 (1994).

14. S. H. Haughton. On a collection of Karroo vertebrate fossils from Tanganyika Territory, *Quart. J. Geol. Soc. London* 88: 634-668 (1932) .

15. S. H. Haughton and A. S. Brink. A bibliographic list of the Reptilia from the Karroo beds of Africa, *Palaeont. Afr.* 2: 1-187 (1954).

16. N. Hotton III . Dicynodonts and their role as primary consumers. In: *The ecology and biology of mammal-like reptiles*. N. Hotton III, P. D. MacLean, J. J. Roth and E. C. Roth (Eds.). pp. 71-82.Smithsonian Institution Press, Washington, D. C. (1986).

17. F. V. Huene. Die Anomodontien des Ruhuhu-Gebeits in der Tübingen Sammlung, *Palaeontogr. Abt. A* 94: 154-184 (1942).

18. T. S. Kemp. Vertebrate localities in the Karroo System of the Luangwa Valley, Zambia, *Nature* 254: 415 (1975).

19. A. W. Keyser. A re-evaluation of the cranial morphology and systematics of some tuskless Anomodontia, *Mem. Geol. Soc. S. Afr.* 82: 81-108 (197).

20. G. M. King. The functional anatomy of a Permian dicynodont, *Phil. Trans. Roy. Soc. London B*, 291: 243-322 (1981).

21. G. M. King. Anomodontia, *Handb. Paläoherp.* 17C: 1-174 (1988).

22. G. M. King. *The dicynodonts a study in paleobiology*. Chapman and Hall, London (1990).

23. J. W. Kitching. Distribution of the Karroo vertebrate fauna, *Mem. Bernard Price. Inst. Palaeont. Res.* 1: 1-131 (1977).

24. S. G. Lucas. The *Dicynodon* biochron and the unity of Late Permian Pangea, *30th Int. Geol. Cong. Abs.* 2: 94 (1996) .

25. I. Metcalfe. Origin and assembly of south-east Asian continental terranes. In: *Gondwana and Tethys*. M. G. Audley-Charles and A. Hallam. (Eds.). pp. 101-118. Geol. Soc. Spec.Paper 37, London (1988).

26. I. Metcalfe. Pre-Cretaceous evolution of SE Asian terranes. In: *Tectonic evolution of Southeast Asia*. R. Hall and D. Blundell. (Eds.). pp. 97-122. Geol. Soc. Spec. Pub. 106, London (1996).

27. E. T. Newton. On some new reptiles from the Elgin Sandstones, *Phil. Trans. Roy. Soc. London B* 184: 431-503 (1893).

28. J. Piveteau. Un thérapsidé d'Indochine remarques sur la notion de continent de Gondwana, *Ann Paléont*. 27: 139-152 (1938).

29. J. Repelin. Sur un fragment de crâne de *Dicynodon* recuelli par H. Counillon dans les environs de Luang-Prabang (Haut-Laos). *Bull. Serv. Geol. Indochine* 12: 1-7 (1923).

30. T. Rowe. The morphology, affinities and age of the dicynodont reptile *Geikia e lginensis*. In: *Aspects of vertebrate history*. L. L. Jacobs. (Ed.). pp. 269-294. Museum of Northern Arizona Press, Flagstaff, Arizona (1980).

31. A. Sun. Permo-Triassic dicynodonts from Turfan, *Mem. Inst. Vert. Paleo. Paleoanthro. Beijing* 10: 53-68 (1973).

32. A. Sun. Two new genera of Dicynodontidae, *Mem. Inst. Vert. Paleo. Paleoanthro. Beijing* 13: 19-25 (1978).

33. P. P. Sushkin. Notes on the pre-Jurassic Tetrapoda from Russia. 1. *Dicynodon amalitzki, Palaeont. Hung.* 1: 323-327 (1926).

34. A. D. Walker The age of the Cutties Hillock Sandstone (Permo-Triassic) of the Elgin area, *Scottish J. Geol.* 9: 177-183 (1973).

35. P. L. Yuan and C. C. Young. On the discovery of a new *Dicynodon* in Sinkiang, *Bull. Geol. Soc. China* 13: 563-573 (1934).

36. Y. Zhu. The discovery of dicynodonts in Daqingshan Mountain, Nei Monggol (Inner Mongolia),*Vert. PalAsiatica* 27: 9-27 (1989).

37. A. M. Ziegler. Phytogeographic patterns and continentalconfigurations during the Permian Period. In: *Paleozoic palaeogeography and biogeography*. W. S. McKerrow and C. R. Scotese. (Eds.). pp. 363-379. Geol. Soc. Mem. 12, London (1990).

Proc. 30ʰ Int'l. Geol. Congr., Vol. 11 , pp. 143-152
Wang Naiwen and J. Remane (Eds)
© VSP 1997

CORRELATION OF THE PERMIAN-TRIASSIC BOUNDARY IN ARCTIC CANADA AND COMPARISON WITH MEISHAN, CHINA

CHARLES M. HENDERSON
Department of Geology and Geophysics, University of Calgary, Calgary, Alberta, CANADA
and
AYMON BAUD
Geological Museum, UNIL-BFSH 2, CH-1015, Lausanne, SWITZERLAND.

ABSTRACT

Correlation of the Permian-Triassic boundary (P-T) is impeded, in part, by the lack of biostratigraphic correlations between tethyan and boreal settings and the presumed unconformity at many P-T sections in the world. Sequence stratigraphic and biostratigraphic data were collected from six sections of the basal part of the Blind Fiord Formation (Confederation Point Member) on northwestern Ellesmere and Axel Heiberg islands; sequence biostratigraphic data are presented in this paper for only the Griesbach Creek and South Otto Fiord sections. These data address some of these impediments, and allow comparisons to be made with the possible Global Stratotype Section and Point (GSSP) at Meishan, China.

Keywords: Permian-Triassic boundary, conodonts, ammonoids, biostratigraphy, Arctic Canada, China

INTRODUCTION
LOCATION

The study area (Fig. 1) is located in the Sverdrup Basin on northwestern Ellesmere Island (Otto Fiord South Section: 81°08'N,084°59'W) and Axel Heiberg Island (Griesbach Creek Section: 80°28'N, 094°30'W). Throughout the Permian and Triassic, the Sverdrup Basin, on the northern margin of Pangea (about 45°N paleolatitude during P-T), was surrounded by a continental land mass to the north and south [5]. Only narrow and probably shallow connections existed with nearby oceans during the Upper Permian, permitting exchange with the global ocean system [1].

GEOLOGY

As a result of sea-level fluctuations the Sverdrup Basin may have been nearly landlocked during the latest Permian; a basin wide unconformity has been interpreted to separate Permian and Triassic rocks in the basin [14]. The base of the Triassic has traditionally been correlated with the base of the Confederation Point Member (CPM) shales of the Blind Fiord Formation; Lower Griesbachian ammonoids are often recovered several metres above the contact. However, recently an uppermost Permian unit comprising shale and spiculitic chert has been interpreted as a transgressive systems tract of an uppermost Permian-Lower Triassic sequence, the maximum flooding surface of which lies within the lower CPM shales [revised herein from 2]. This sequence is the first of nine regional third-order transgressive-regressive sequences recognized in the Triassic succession of the

144

Sverdrup Basin and is dated as Upper Changhsingian to Upper Griesbachian [revised herein from 4].

This sequence begins with 10 cm of platy, glauconitic, and spiculitic siltstone which is overlain by 10 cm of spiculitic, sandy chert. These units are overlain by 5 cm of yellow clay, which may be a paleosol (suggesting that this spiculitic unit may be a slightly older sequence with a minor unconformity at the top) or a more recent weathering feature, and then 25 cm of greenish-grey clay. Above the clays, the sequence includes dark grey to black, laminated silty shale with pyrite, followed by black, fissile, pyritic shale (the maximum flooding surface or zone; MFS), which is overlain by dark grey to black, laminated silty shale. The succession is then capped with resistant, ripple cross-laminated siltstone with concretions at Otto Fiord South or by resistant very fine grained sandstone at Griesbach Creek (Fig. 2).

Figure 1. Sections studied for this paper are from the Canadian Arctic Archipelago on northwestern Ellesmere and Axel Heiberg Islands. The position of the Sverdrup Basin margin is indicated (dotted line), as well as the location of the Otto Fiord South (OFS) and Griesbach Creek (Gb) sections.

BIOSTRATIGRAPHY
INTRODUCTION

Conodont samples and macrofossil collections (bivalves and ammonoids) were collected from six sections in the Canadian Arctic. This paper documents the recovered faunas from the two most complete sections at Griesbach Creek and Otto Fiord South (Fig. 2). A longer paper in preparation by the authors, provides more details regarding the recovered fossils, taxonomy, and measurements for all of the sections.

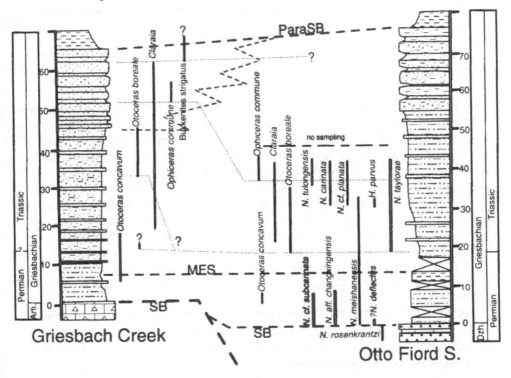

Figure 2. Lithostratigraphic and biostratigraphic correlation of the Griesbach Creek and Otto Fiord South sections. The range of key molluscan and conodont species are indicated for each section. Key sequence stratigraphic boundaries (sequence boundaries (SB), maximum flooding surface (MFS), and parasequence boundaries (ParaSB)) are indicated by dashed lines. Ammonoid biozonal boundaries (including the *boreale*, *commune*, and *strigatus* zones) are indicated by dotted lines. The sequence stratigraphic and biozonal boundaries serve as datums for comparison to the ranges of conodont taxa. No samples were collected from above 45 m at Otto Fiord South. Lithology can be discerned from text. Arti.=Artinskian Great Bear Cape Fm.; Dzh.=?Dzhulfian upper Degerbols/Van Hauen fms.; Griesbachian=Confederation Point Mbr. Blind Fiord Fm.

GRIESBACH CREEK SECTION

The basal Blind Fiord Formation at the Griesbach Creek section, the type area for the Griesbachian, rests unconformably upon cherty limestones of the Artinskian Great Bear Cape Formation. This area was high for a considerable part of the upper Permian. The MFS for this section is interpreted at about 8 m above the base. All conodont samples from this section were barren, which may be attributed to biofacies as the Griesbach Creek section is more proximal compared to Otto Fiord South. However, ammonoids and bivalves are exceptionally abundant at this locality; these have previously been reported by Tozer [15]. Additional specimens were collected by the authors. *Otoceras concavum*, *Otoceras boreale*, *Ophiceras* spp. (including *O. commune*), and *Bukkenites strigatus* occur in succession. According to Tozer [15], a gap of about 15 m separates the last *Otoceras concavum* from the first *O. boreale*. However, collections by the authors questionably suggest that the two species may overlap (Fig. 2). The bivalve, *Claraia* was found immediately above *O. concavum* and overlaps the range of *Otoceras boreale* and *Ophiceras commune*. These molluscan fossils provide biozonal datums for comparison to molluscan and conodont distribution at Otto Fiord South.

OTTO FIORD SOUTH

The basal Blind Fiord Formation at Otto Fiord South rests disconformably (nearly conformable?) on a thin spiculitic shale and sandy chert unit, which is underlain by chert and carbonate of the Degerbols Formation. The upper Degerbols and thin spiculitic shale both contain *Neogondolella rosenkrantzi*. A thin yellow clay overlies the spiculite unit. This is in turn overlain by pyritic, laminated black shale and siltstone that may have been deposited under anoxic or dysaerobic conditions [19, 20]; supporting evidence includes the lack of trace fossils and the lack of a benthic biota other than a couple of inarticulate brachiopod fragments and ammodiscid foraminifers. However, pelagic ammonoids and conodonts were recovered from this interval. The MFS for this section is interpreted at about 14 m. The conodont fauna can be subdivided into three components:
1. BELOW THE MFS, comprising *Neogondolella* sp.cf. *subcarinata*, *N.* sp.aff. *changxingensis* (possibly similar to *N. orchardi* [7]), *N. meishanensis*, and possibly *N. deflectus*;
2. ASSOCIATED WITH *OTOCERAS BOREALE*, comprising *N. taylorae*, *N. carinata*, *N.* sp.cf. *planata*, *N. tulongensis*, *N. meishanensis*, and *Hindeodus parvus*; and
3. ASSOCIATED WITH *OPHICERAS COMMUNE*, comprising *N. taylorae*, *N. carinata*, *N. tulongensis*, and *N.* sp.cf. *planata*.

CONODONT FAUNA AND COMPARISON WITH MEISHAN, CHINA
INTRODUCTION

For the purpose of comparison the pertinent conodonts can be subdivided into four assemblages (A-D, Fig. 3). Some taxonomic remarks are given, but these are kept to a minimum. The authors are preparing a longer paper that will discuss the data at all Arctic sections and provide taxonomic description and illustration. Some taxonomic comparisons discussed below are not exact (e.g. *N.* sp.cf. *subcarinata*) but, given the variation within most conodont populations and the widely separated regions, they are considered reasonable for correlation. Population studies are needed that illustrate all variants so that changes in genetic composition (phenotype) can be assessed geographically and in different environments. Some variants will be better suited (more successful) to different regions thus altering selection pressure and gene flow between populations in such widely disparate regions as the South China tethys and the boreal Sverdrup Basin.

ASSEMBLAGE A

This assemblage, which includes *N. rosenkrantzi*, was only recovered from the Otto Fiord section and was briefly discussed above for the Otto Fiord South section. This species may range as high as Lower Dzhulfian [9], but the occurrences here could be even younger if the Sverdrup Basin was restricted from migration and competition of other neogondolellid species during the uppermost Permian as a result of lowstand.

ASSEMBLAGE B

This assemblage occurs in most of the Changhsingian at Meishan and includes *Neogondolella subcarinata, N. changxingensis, N. deflecta,* and *Hindeodus typicalis* [23]. Taxonomic nomenclature varies somewhat in various publications for the neogondolellid species. This assemblage and apparently the chronostratigraphic interval that they represent are not present in the Sverdrup Basin of the Canadian Arctic.

ASSEMBLAGE C

This assemblage occurs in the lower 14-18 metres at the Otto Fiord Section and probably in beds 24e to 26 at Meishan (a 20 cm thick interval). This assemblage at Otto Fiord includes *Neogondolella* sp.cf. *subcarinata* (with high, fused posterior carina), *N.* sp.aff. *changxingensis* (similar to *N. orchardi* [12], with pointed posterior and low, rather discrete posterior carina [alpha morphotype of 10]), *N. meishanensis* (with low, discrete posterior carina and round posterior margin), and questionable *N. deflectus* (may be gerontic form of associated species). This fauna compares questionably with the fauna in bed 24e at Meishan where *N. subcarinata* (bed 24e) occurs; the species at Otto Fiord are similar, compared to, and may in fact be *N. subcarinata* (see introduction for this section). The fauna also compares favourably with the fauna in beds 25-26 where *N. changxingensis, N. deflecta,* and *N. meishanensis* occur. However, species of *Hindeodus,* including *H. typicalis* and *H. latidentatus* were also recovered from this interval at Meishan. Some specimens of *H. latidentatus* from bed 25 (Plate II, Fig. 12 of [23]; same specimen also illustrated in Plate II, Fig. 4 of [7]) are similar to *Hindeodus parvus* (Mei [12] identifies them as *H. parvus*). Kozur [7] indicates that this is a transitional form. We agree that this form appears transitional; the specimen has the *parvus* cusp and the posterior denticulation of *latidentatus.* More work is clearly needed on these hindeodid species before a lineage can be demonstrated [7, 8, 12]. This assemblage may also correlate with those recovered from the Changhsingian bed 19 (8-17 cm) at Selong [12] despite differing opinions regarding the significance of this bed [6, 13].

ASSEMBLAGE D

This assemblage occurs from 18 to 36 metres at the Otto Fiord Section and is associated with *Otoceras boreale* and *Claraia.* Conodonts recovered include *N. tulongensis, N. carinata, N. taylorae, N. meishanensis, N.* sp.cf. *planata,* and *Hindeodus parvus. Neogondolella taylorae* begins just above the MFS, but the other named species occurs several metres above the MFS. This compares well with beds 27c-30 at Meishan where *H. parvus, N. carinata,* and *N. planata* all occur, however, *Neogondolella meishanensis* is

148

not present, suggesting an extended range at Otto Fiord. Bed 27 occurs immediately above a transgressive surface well below the supposed MFS at the base of bed 35 [13]. However, the extremely condensed nature of this outcrop must make sequence stratigraphic correlation difficult, and it is possible that higher order parasequences, as at Otto Fiord, may not be discerned. The TS at Meishan appears to roughly coincide with the MFS at Otto Fiord. This faunal assemblage also appears to correlate with bed 20 (10-30 cm) at Selong which is associated with *Otoceras latilobatum*. At Selong, *Hindeodus parvus* and *Neogondolella taylorae* appear together [12, 13]. The wide separation of appearances of these two species at Otto Fiord suggests that some biofacies control may be exhibited on the distribution of *Hindeodus parvus*.

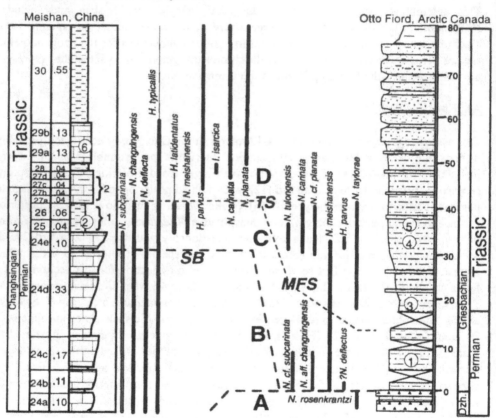

Figure 3. Comparison of conodonts from the Otto Fiord Section and from Meishan, China (data for the latter modified from [23]). Four conodont faunas or assemblages are indicated (A-D; A is only present at OFS, and B only at Meishan. The maximum flooding surface at Otto Fiord South is correlated with the transgressive surface (TS from [23]) at Meishan. Thicknesses are in metres; note the very different scales which reflects the condensation at the Meishan section. The position of key ammonoids are indicated in circles for reference (1. *Otoceras concavum?*, 2. *Otoceras?*, 3. first *Otoceras boreale*, 4. last *Otoceras boreale*, 5. *Ophiceras commune*, 6. *Ophiceras*).

IMPLICATIONS REGARDING PERMIAN-TRIASSIC BOUNDARY

Data recently gathered from the Canadian Arctic and discussed herein by the authors have important implications regarding the Permian-Triassic boundary. These data provide a correlation between tethyan and boreal settings and between areas with very different sedimentation rates. These implications are summarized below.

1. Biostratigraphic data (Figs. 2,3) suggest that basal Griesbachian strata in the Canadian Arctic (lower Confederation Point Mbr. of Blind Fiord Fm.) correlate with the uppermost Changhsingian at Meishan (?beds 24e-27b). This assumes that the transition beds (25-27b), with which the suggested correlation is strongest, will be referred to as Changhsingian [17] if and when the base of bed 27c is chosen as the GSSP (see point 3). This indicates that the transgression at the base of the Blind Fiord Formation begins in latest Permian and not in the earliest Triassic. This also indicates that the anoxic interval in the Arctic begins in the latest Permian and could therefore be a contributing factor towards the extinction [19, 20], a problem if it occurred in the early Triassic after the major extinctions.

2. Tim Tozer [16] has long repeated the opinion that the base of the Triassic should be defined by *Otoceras*. Our most recent work in the Canadian Arctic confirms that species of *Otoceras* may be a good index (at least an auxiliary index) for the Permian-Triassic boundary. At Otto Fiord South, the appearance of *Neogondolella taylorae* coincides with the appearance of *Otoceras boreale*, and *Hindeodus parvus* occurs near the top of the zone of *Otoceras boreale*. Assuming that the correlations presented in point 1 are correct, uppermost Permian conodonts are associated with *Otoceras concavum* at Otto Fiord South. At Selong, *Otoceras latilobatum*, *Neogondolella taylorae*, and *Hindeodus parvus* appear simultaneously [12, 13]. There is no reason to believe that the two *Otoceras* zones defined at Otto Fiord (*concavum* and *boreale* Zones) must be exactly correlative to the two *Otoceras* zones at Selong (*latilobatum* and *woodwardi*) as has been suggested [3] or repeated [13], since they are defined by different species. It appears that *Otoceras boreale* as well as *O. latilobatum* could be used as auxiliary indices for the base of the Triassic [21]. This would place the boundary in the Canadian Arctic within the lineage of *Otoceras concavum* to *O. boreale*, which makes it even more attractive as a boundary indicator.

3. Yin Hongfu and others [21, 18, 25] have recommended the introduction of *Hindeodus parvus* at the base of bed 27c at Meishan as the P-T GSSP. Our work in the Arctic, in general, supports this position by demonstrating the presence of this key species in the Sverdrup Basin, but it also indicates problems with respect to the first appearance position of this taxon. In fact, one of us (AB) does not agree with the recommended proposal. The introduction of *H. parvus* at the base of bed 27c at Meishan is apparently in accordance with the international stratigraphic code, although questionably with respect to condensation, and there is considerable agreement within the P-T Boundary Working Group. It may be time to make a formal decision and get on with the study of the nature of events at and around this boundary. However, it is necessary to point out some of the problems with this GSSP, keeping in mind that no section or point will be perfect. These problems include the condensation noted at the tethyan sections, possible biofacies problems especially with respect to the first appearance of *H. parvus* (although the probability that the first appearance of any taxon is synchronous in all sections is virtually zero), the need for auxiliary indices, and the position with respect to sequence stratigraphic boundaries. These are addressed in points 4-7 that follow.

4. The tethyan sections exhibit low sedimentation rates and facies changes occur close to the boundary although apparently not right at the boundary (beds 27a-d are a single facies, but the proposed GSSP occurs only 8 cm above a facies change). The boundary interval at Otto Fiord is on the order of 40 metres whereas at Meishan it is about 40 centimetres (similar at Selong), a condensation factor of 100X. There are no specifications in the code regarding condensation, but extremely condensed sections may have the effect of consolidating separate biotic and abiotic events; an undesirable result given the high resolution necessary for boundary studies.

5. The first appearance of *H. parvus* is considered a good GSSP index by one of us (CMH) because it has distinctive morphology and apparently occurs within a lineage of *Hindeodus* species [7, 12]. However, we agree with Mei [12] that this lineage has not been clearly defined at any one location. Furthermore, forms transitional between *H. latidentatus* and *H. parvus* in bed 25 could easily be mistaken for *H. parvus*. Population studies of these species are clearly needed. However, in the Canadian Arctic, *H. parvus* is exceptionally rare. Furthermore, it is possible that the specimens from Otto Fiord South do not represent the true first appearance as they occur considerably above the appearance of *Neogondolella taylorae*; elsewhere they coincide [13]. The appearance, and even more likely the abundance, may be controlled by biofacies. Kozur [7] suggests that *H. parvus* occurs in both shallow and deep water facies, but older species of *Hindeodus* (Upper Paleozoic) are nearly always indices for shallow, and often shallow restricted marine facies. Other "deeper water" conodont species like *Neogondolella* may be useful as an index or auxiliary index for the GSSP.

6. *Neogondolella taylorae* [13] may be a good index or auxiliary index for the Permian-Triassic boundary. It appears coincident with *Otoceras latilobatum*, *Hindeodus parvus* [13] and *Neospathodus primitivus* [12], a new and oldest species for this genus at Selong and with *Otoceras boreale* in Arctic Canada. Neogondolellid species are apparently recovered in greater abundance at more locations than are hindeodid species. However, there is considerable disagreement regarding taxonomy of the neogondolellids because these species tend to have somewhat "plastic" morphology. Recent taxonomic work seems to be coming closer to a consensus [10, 12, 13]. The comparison of ranges between *Neogondolella taylorae* and *Hindeodus parvus* could help reconcile potential biofacies control on appearance.

7. Finally, a number of workers suggest that the P-T boundary should be at the base of the transitional beds (base of bed 25) or at the conformable part of the Sequence Boundary [11, 12]. According to Zhang et al [24] the Sequence Boundary occurs at the base of bed 24e; this coincides with the extinction of many tethyan biotic indices [22]. Whereas the sequence boundary in some respects may be the best natural boundary [12], it will also be the least correlatable, as the boundary will be unconformable at most sites. According to the stratigraphic code, boundary stratotypes should be identified so as to permit widespread tracing of a synchronous horizon. It therefore seems appropriate to place the GSSP some distance above the SB so that better correlation may be achieved. However, the base of bed 25 at Meishan, which is correlated with the base of bed 20 at Selong [12], is recognized as a potential GSSP by many workers [12] and is above the sequence boundary. Moving a little higher above the SB can only improve the correlation potential of the

GSSP. Mei [11], in a strong argument, suggests that the conformable part of the SB makes the best natural boundary and that it can be easily recognized and traced in outcrop sections without biostratigraphy, because the unconformity will exhibit major lithologic and biotic shifts. However, the fact that a GSSP above the SB does not occur at a major lithologic break should not be considered a problem. The major lithologic break can still be correlated as a diachronous lithostratigraphic unit for mapping purposes, with the knowledge that the GSSP is some distance above; the P-T GSSP should not coincide with a diachronous unit and cannot be considered useful if it can only be recognized at one location.

SUMMARY

Conodonts recovered from the basal Confederation Point Member of the Blind Fiord Formation correlate with the latest Changhsingian and transition beds at Meishan, confirming that a major transgression began during the latest Permian, rather than during the earliest Triassic. The results suggest that *Hindeodus parvus*, *Neogondolella taylorae*, or *Otoceras boreale* are potential indices for the GSSP. Problems associated with the possible GSSP position of bed 27c at Meishan were addressed. The Arctic sections may make suitable supplementary (Boreal) references despite their difficult accessibility, as they occur in thick sections with continuous deposition and without major facies changes, as well as containing, although rarely in abundance, conodont, ammonoid, and bivalve index fossils.

ACKNOWLEDGEMENTS

The authors thank Dr. Mike Orchard of GSC Vancouver and Dr. Benoit Beauchamp of GSC Calgary for reviewing the manuscript and providing useful suggestions. Conodont samples were processed by Lorraine Bloom of the University of Calgary. We are also grateful to the Geological Survey of Canada which provided logistical support in the Canadian Arctic and to support from a NSERC Research Grant to CMH.

REFERENCES

1. Beauchamp, B., Oldershaw, A.E., and Krouse, H.R. Upper Carboniferous to Upper Permian ^{13}C enriched primary carbonates in the Sverdrup Basin, Canadian Arctic; comparisons to coeval western North American ocean margins. *Chemical Geology (Isotope Geoscience Section)*, 65: 391-413, (1987).
2. Beauchamp, B., Mayr, U., Harrison, J.C., and Desrochers, A. Uppermost Permian stratigraphy (Unit P6), Hvitland Peninsula and adjacent areas, northwestern Ellesmere Island, Arctic Canada. *In* Current Research, 1995-B, *Geological Survey of Canada*. 65-70, (1995).
3. Dagis, A.S. and Dagis, A.A. Biostratigraphy of the lowermost Triassic and the boundary between Paleozoic an Mesozoic. *Societa Geologica Italiana*, Mem. 34: 313-320, (1988).
4. Embry, A.F. Triassic Sea-level changes: Evidence from the Canadian Arctic Archipelago. In Wilgus, Cheryl K. et al. (eds), *SEPM Special vol.* 42, Sea-level changes: An Integrated Approach. 249-259, (1988).
5. Embry, A.F. Mesozoic history of the Arctic Islands; Chapter 14. *In* Geology of the Innuitian Orogen and Arctic Platform of Canada and Greenland, H.P. Trettin (ed.). *Geological Survey of Canada* no. 3, v. E: 371-434, (1991).
6. Jin Yugan, Shen Shuzhong, Zhu Zili, Mei Shilong, and Wang Wei. The Selong section, Candidate of the global stratotype section and point of the Permian-Triassic Boundary. *In* The Palaeozoic-Mesozoic boundary candidates of global stratotype section and point of the Permian-Triassic Boundary (Yin Hongfu, ed.), *China University of Geosciences Press*: 127-139, (1996).

7. Kozur, H. Some remarks to the conodonts *Hindeodus* and *Isarcicella* in the latest Permian and earliest Triassic. *Palaeoworld, Academia Sinica*. **6**: 64-80, (1995).

8. Kozur, H.W., Ramovs, A., Cheng-yuan Wang, Zakharov, Y.D. The importance of *Hindeodus parvus* (Conodonta) for the definition of the Permian-Triassic boundary and evaluation of the proposed sections for a global stratotype section and point (GSSP) for the base of the Triassic. *Geologija, Ljubljana*, **37,38**: 173-213, (1994/95).

9. Kozur, H. and Mostler, H. Guadalupian (Middle Permian) conodonts of sponge-bearing limestones from the margins of the Delaware Basin, West Texas. *Geol. Croat., Zagreb*. **48/2**: 107-128, (1995).

10. Krystyn, L. and Orchard, M.J. Lowermost Triassic ammonoid and conodont biostratigraphy of Spiti, India. *Albertiana*. In press (1996).

11. Mei, Shilong. The best natural boundary: a new concept developed by combining the sequence boundary with the GSSP. *Acta Geologica Sinica*. **9**: 98-106, (1996).

12. Mei, Shilong. Restudy of conodonts from the Permian-Triassic boundary beds at Selong and Meishan and the natural Permian-Triassic Boundary. Palaeontology and Stratigraphy (Wang Hongzhen and Wang Xunlian; eds.), *China University of Geosciences Press*: 141-148, (1996).

13. Orchard, M.J., Nassichuk, W.W., and Rui Lin. Conodonts from the Lower Griesbachian *Otoceras latilobatum* bed of Selong, Tibet and the position of the Permian-Triassic boundary. *In* Pangea: Global Environments and Resources, *Canadian Society of Petroleum Geologists*. **Memoir 17**: 823-843 (1994).

14. Thorsteinsson, R. Carboniferous and Permian stratigraphy of Axel Heiberg Island and western Ellesmere Island, Canadian Arctic Archipelago. *Geological Survey of Canada, Bulletin* **224**: 115 pp., (1974).

15. Tozer, E.T. A standard for Triassic Time. *Geological Survey of Canada, Bulletin* **156**: 103 pp., (1967).

16. Tozer, E.T. Definition of the Permian-Triassic (P-T) boundary: the question of the age of the *Otoceras* beds. *Mem. Soc. Geol. Italiana*, **36**: 291-302, (1986).

17. Wang, Cheng-yuan. A conodont-based high-resolution eventostratigraphy and biostratigraphy for the Permian-Triassic boundaries in South China. *Palaeoworld, Academia Sinica*, **4**, 234-248, (1994).

18. Wang, Cheng-yuan. Conodonts of Permian-Triassic boundary beds and biostratigraphic boundary. *Acta Palaeontologica Sinica*. **34**: 129-151, (1995).

19. Wignall, P.B. and Hallam, A. Griesbachian (Earliest Triassic) palaeoenvironmental changes in the Salt Range, Pakistan and southeast China and their bearing on the Permo-Triassic mass extinction. *Palaeogeography, Palaeoclimatology, Palaeoecology*. **102**: 215-237, (1993).

20. Wignall, P.B. and Twitchett, R.J. Oceanic anoxia and the end Permian mass extinction. *Science*. **272**: 1155-1158, (1996).

21. Yin Hongfu, Wu Shunbao, Ding Meihua, Zhang Kexin, Tong Jinnan, Yang Fengqing, and Lai Xulong. The Meishan section, candidate of the Global Stratotype Section and Point of the Permian-Triassic Boundary. *In* The Palaeozoic-Mesozoic boundary candidates of global stratotype section and point of the Permian-Triassic Boundary (Yin Hongfu, ed.), *China University of Geosciences Press*: 31-48, (1996).

22. Yin Hongfu and Zhang Kexin. Eventostratigraphy of the Permian-Triassic boundary at Meishan section, South China. *In* The Palaeozoic-Mesozoic boundary candidates of global stratotype section and point of the Permian-Triassic Boundary (Yin Hongfu, ed.), *China University of Geosciences Press*: 84-96, (1996).

23. Zhang Kexin, Lai Xulong, Ding Meihua, and Liu Jinhua. Conodont Sequence and its global correlation of Permian-Triassic Boundary in Meishan section, Changxing, Zhejiang Province. *Earth Science-Journal of China University of Geosciences*. **20**, **6**: 669-676, (1995).

24. Zhang Kexin, Tong Jinnan, Yin Hongfu, and Wu Shunbao. Sequence stratigraphy near the Permian-Triassic boundary at Meishan section, South China. *In* The Palaeozoic-Mesozoic boundary candidates of global stratotype section and point of the Permian-Triassic Boundary (Yin Hongfu, ed.), *China University of Geosciences Press*: 72-83, (1996).

25. Zunyi Yang, Jinzhang Sheng, and Hongfu Yin. The Permian-Triassic boundary: the global stratotype section and point (GSSP). *Episodes*, **18**: 49-53, (1995).

Proc. 30ᵗʰ Int'l. Geol. Congr., Vol. 11 , pp. 153-162
Wang Naiwen and J. Remane (Eds)
© VSP 1997

Conodont Sequences and Their Lineages in the Permian-Triassic Boundary Strata at the Meishan Section, South China

DING MEIHUA,   LAI XULONG,   ZHANG KEXIN

Faculty of Earth Sciences, China University of Geosciences, Wuhan 430074, China

Abstract

This paper deals in detail with the conodont sequences and the evolution of *Clarkina* lineage and *Hindeodus--Isarcicella* lineage on the basis of the Upper Permian -Lower Triassic Conodonts from the Meishan section, Changxing, Zhejiang Province, South China, Which is proposed as a Candidate of the Global Stratotype Section and Point of the Permian-Triassic boundary. Three Upper Permian-Lower Triassic Conodont Zones are defined in ascending order, as follows. *Clarkina changxingensis* Zone, *Hindeodus Parvus* Zone, and *Isarcicella isarcica* Zone. The *Clarkina changxingensis* Zone is subdivided into three faunas, the *Clarkina changxingensis-C. dleflecta-C. subcarinata* fauna, the *Hindeodus latidentatus-Clarkina meishanensis* fauna and the *Hindeodus typicalis* fauna. It is very important that the conodont sequences established in the Permian-Triassic boundary strata at the Meishan section make a complete zonation and correlation over the world. This significantly helps the precise definition of the Permian-Triassic boundary at the Meishan section. The *Clarkina* stock at section D of Meishan, Changxing, appeared only in the Upper Permian and represented the descendant of the Eurasian stock. It is mainly Composed of *Clarkina orientalis* (Barskov and Koroleva), *C. subcarinata* (Sweet), *C. wangi* (Zhang), *C. changxingensis* (Wang and Wang), *C. deflecta* (Wang and Wang), *C. carinata* (Clark) and *C. dicerocarinata* (Wang and Wang). *Clarkina orientalis* is the ancestor of *C. subcarinata*. *C. subarinata* diverged into three branches, i.e., *C. Changxingensis*, *C. deflecta* and *C. carinata*. *C. deflecta* further developed into *C. dicerocarinata*. The evolution of *Clarkina wangi* is unsettled. The evolution of *Hindeodus-Isarcilella* stock at the Meishan section, Changxing, is *Hindeodus latidentatus-H. parvus-Isarcicella turgida-I. isarcica*.

Keywords: conodont, sequence, lineage, Permian, Triassic, boundary, Meishan, South China

INTRODUCTION

Since Yin *et al.* (1988) recommended the first appearance of *Hindeodus parvus* as the base of the global Triassic, this proposal has been received by more and more stratigraphers and paleontologists over the world. Thus, the study of the conodont faunas near the Permian-Triassic boundary becomes more and more important. The conodont biostratigraphy of the Upper Permian to the Lower Triassic at the Meishan section, Changxing, has been studied in detail by many scholars in recent years [16, 28, 10, 20, 25, 21, 14, 15, 24, 6, 27, 26, 3, 17] .In order to explore the conodont-bearing beds and establish the detailed conodont sequences and lineages near the Permian-Triassic boundary strata at the Meishan section, vigorous works have been carried out at the Meishan section by our conodont group. Zhang, who is one of the authors, and two graduate students from 1993 to 1995, collected 120 samples of conodonts bed by bed from Bed 24e to Bed 29 near the Permian-Triassic boundary, which were distributed into 7 quarries from east to west at Meishan. Average collection interval is 6 cm, and the interval collected from Bed 27a to Bed 28 is 4 cm, We have obtained over 1500 conodont elements from these samples. It is most important that *Isarcicella isarcica* at Bed 28, section A, and *Hindeodus latidentatus* at Bed 25, section B, are first discovered at Meishan [6,27] . Thus, new conodont data from the Meishan section make it is possible to establish an entire conodont sequence   and to discuss the conodont lineages near the Permian-Triassic boundary at the Meishan section.

CONODONT ZONES

According to the conodont occurrences in the 7 sections of Meishan, the conodonts occurring from Bed 29 near the Permian-Triassic boundary strata are divided into four conodont zones, in ascending order, as follows (Fig. 1):

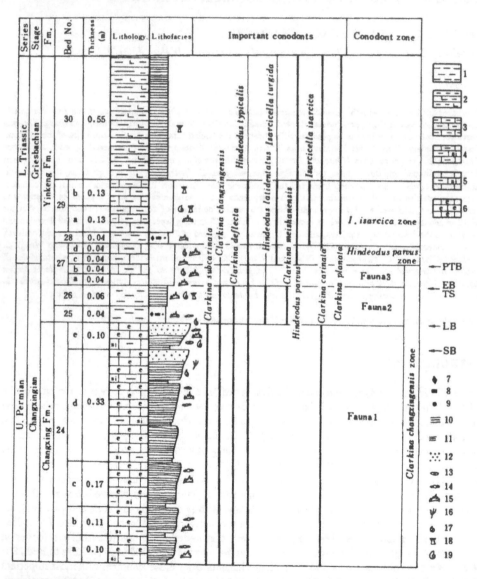

Figure 1. Distribution of main conodonts of the Upper Permian- Lower Triassic at Meishan section, Changxing County
1. clay; 2. calcareous mudrock; 3. marl; 4. argellaceous micrite; 5. siliceous micrite; 6. bioclastic micrite; 7. high quartz; 8. zircon; 9. microsphaerules; 10. horizontal bedding; 11. wavy bedding; 12. convolute bedding; 13. fusulinids; 14. foraminifera; 15. conodonts; 16. calcareous algae; 17. brachiopods; 18. bivalves; 19. ammonoids.

Clarkina changxingensis Zone

Clarkina changxingensis (Wang and Wang) ranges from Bed 24e to Bed 27b near the Permian-Triassic boundary strata. It extends downward into the middle part of the lower member of the Changxing Formation. The base of the *Clarkina changxingensis* Zone is marked by the appearance of *C. changxingensis* (Wang and Wang) and its top is marked by the appearance of *Hindeodus parvus* (Kozur and Pjatakova). *C. changxingensis* is rare in the lower member of the Changxing Formation, but it is abundantly represented in the upper member. Wang and Wang[16] divided the Changxing Formation into two conodont zones, *Clarkina changxingensis-C. deflecta* Zone and *C. subcarinata-C. wangi* Zone. These zones are acme zones. The boundary between these zones is not clear, and they apparently overlap. Therefore, the definition of these zones established by Wang and Wang[16] was revised by Ding[2]. Ding[2] suggested that the Changxing Formation was divided into two zones, *Clarkina changxingensis* Zone and *C. subcarinata* Zone. The zone contains three faunas from Bed 24e to Bed 27b near the Permian-Triassic boundary, in ascending order as follows:

Fauna 1 The Fauna 1 or *Clarkina changxingensis-C. deflecta-C. subcarinata* fauna consists mainly of *Clarkina changxingensis* (Wang and Wang), *C. deflecta* (Wang and Wang) and *C. subcarinata* (Sweet).The fauna has been recognized in Bed 24e and is marked by the abundance of *Clarkina changxingensis* and terminates with the first appearance of *Hindeodus latidentatus* (Kozur, Mostler and Rahimi-Yazd). This fauna has a high diversity, containing 12 conodont species. The fauna co-occurs with the ammonoid *Rotodiscoceras* sp. and the fusulinid *Palaeofusulina* sp. at the Meishan section.

Fauna 2 The Fauna 2 or *Hindeodus latidentatus-Clarkina meishanensis* fauna occurred within the boundary clay rocks (Bed 25 and Bed 26) of the Permian-Triassic. The fauna began with the first appearance of *H. latidentatus* (Kozur, Mostler and Rahimi-Yazd) and *Clarkina meishanensis* Zhang, Lai, Ding, Wu and Liu and terminates with the disappearance of *Clarkina meishanensis*. The abundance of Fauna 2 is lower than Fauna 1, but its diversity is still high containing 9 conodont species. *Clarkina changxingensis* (Wang and Wang) and *C. deflecta* (Wang and Wang) are still predominant. The beds yielding Fauna 2 are equal to the lower part of the "*Otoceras* bed" described by Zhao *et al.* [28], the "mixed bed 1" recommended by Sheng *et al.*[10], the "Boundary bed" by Wang[14] and the lower part of the "transitional bed" by Yin *et al.* [22,24]. Fauna 2 co-occurs with *Otoceras?* sp., and *Hypophiceras* spp. in Bed 26 (Black clay), which have been traditionally considered to be Early Triassic. However, their origination ages are now being questioned[21,19]. *Otoceras?* sp. and *Hypophiceras* spp. co-occur with several typical Permian fossils, for example, *Pseudogastrioceras*, *Clarkina changxingensis*, *C. deflecta* and Permian brachiopods. Yang Shouren[18] reported that the conodont elements which occurred from the *Hypophiceras* bed in the base of the Lower Triassic belong to the typical elements of the Changxingian of the Late Permian at Dalishan, Zhenjiang City, Jiangsu Province. The conodonts from the Dalishan section have an identical age with Fauna 2. Kozur orally reported during the PTBWG meeting that from Greenland samples sent to Prof. Sweet, he has found early Changxingian *Clarkina subcarinata and C. orientalis* in the lower *Hypophiceras* bed, late Changxingian *Hindeodus latidentatus* in the *Otoceras boreale* bed, and *H. parvus* in the *Ophiceras* bed (Permian-Triassic Boundary Working Group, Newsletter No. 2, 1993). It suggests further that the lower part the *Otoceras* bed and the *Hypophiceras* bed may belong to the top of the Changxingian, and not to the Lower Triassic.

Fauna 3 The Fauna 3 or *Hindeodus typicalis* fauna occurred from Bed 27a to Bed 27b. The fauna begins with the mass extinction of the Permian *Clarkina*, including *Clarkina deflecta* (Wang and Wang), *C. orientalis* (Barskov and Koroleva) and *C. meishanensis* Zhang, Lai, Ding,

Wu and Liu and terminates with the first appearance of *Hindeodus parvus* (Kozur and Pjatakova) at the base of the Lower Triassic. The fauna is marked by the abundance of *Hindeodus typicalis* (Sweet) and a few *Ellisonia* sp.. There is generally a bed with a thickness from several centimeters to several meters below the first appearance of *Hindeodus parvus* (Kozur and Pjatakova) in South China and many places of the world. Fauna 3 is marked by low abundance and low diversity, and represents the evolutionary ebb of organisms between the mass extinction and the organism recovery during the Permian-Triassic alternation. Sweet [12,13] established a *Hindeodus typicalis* Zone in Kashmir and the Salt Range, Pakistan as the first conodont zone in the Lower Triassic. Matsuda[7] subdivided the *Hindeodus minutus* Zone, *Hindeodus parvus* Zone and *Isarcicella isarcia* Zone in ascending order. Matsuda[7] suggested that the *Hindeodus minutus* Zone and *Hindeodus parvus* Zone corresponded to the lower and upper part of the *Otoceras woodwardi* Zone, respectively, and the *Isarcicella isarcica* Zone was assigned to the lower part of the *Ophiceras tibeticum* Subzone. The authors suggest that *Hindeodus latidentatus-Clarkina meishanensis* fauna and *Hindeodus typicalis* Fauna correspond to the *Hindeodus minutus* Zone of Matsuda [7] and correlate it with the lower part of the *Otoceras woodwardi* Zone; the *Hindeodus parvus* Zone corresponds to the upper part of the *Otoceras woodwardi* Zone and the *Isarcicella isarcica* Zone correlates with the lower part of the *Ophiceras* Zone (see below). The bed containing the Fauna 3 at the Meishan section corresponds to the lower part of the "mixed bed 2" of Sheng [10,11], the lower part of the "Boundary bed 2"[14] and the lower part of the "Upper transitional bed " [22,24].

Hindeodus parvus Zone

This zone ranges from bed 27c to bed 27d in the Lower Triassic. The base of this zone is marked by the appearance of *Hindeodus parvus* (Kozur and Pjatakova) . The top of the zone is marked by the appearance of *Isarcicella isarcica* (Huckriede). Genus and species of the zone are rather monotonous. Wang [14] reported that Upper Permian *Clarkina changxingensis* (Wang and Wang) and *Hindeodus julfensis* (Sweet) were found from Bed 882-3 and Bed 882-4 (=Bed 27c and Bed 27d) at the Zhongxin Dadui section, Changxing. This shows that the relic *Clarkina changxingensis* (Wang and Wang) and *Hindeodus julfensis* (Sweet) may extend upward to the base of the Lower Triassic and co-occur with *Hindeodus parvus* (Kozur and Pjatakova), and the conodont fauna of the zone per se has a mixed character of Permian and Triassic types. This zone corresponds to the upper part of the "mixed bed 2 "of Sheng *et al.*[10,11], the upper part of the "Boundary bed 2 "of Wang [14] and the upper part of the "Upper transitional bed "of Yin *et al.*[22, 24]. *Hindeodus parvus* (Kozur and Pjatakova) is widespread and belongs to early Early Griesbachian.

Isarcicella isarica Zone

This zone ranges from Bed 28 to Bed 29. This biostratigraphic unit is coextensive with the range of the nominate species. The base of this zone is characterized by the first appearance of *Isarcicella isarcica* (Huckriede), but its top is unsettled. The nominate species is found for the first time at Bed 28 of section A of Meishan [6, 27]. At Bed 29a, the nominate species is not found, but *Isarcicella turgida* (Kozur, Mostler and Rahimi-Yazd) occurred from Bed 29a at section D, and *Clarkina planata* (Clark) is found in Bed 29a ,too [11]. In concrete cases, the appearance of *Isarcicella turgida* (Kozur, Mostler and Rahimi-Yazd) may be either later [5] or earlier [9] than *I.isarcica* (Huckriede). It is generally believed that the overlying horizon containing *I. turgida* (Kozur, Mostler and Rahimi-Yazd) still belongs to the *I. isarcica* Zone, even if *I. isarcica* is not found in the horizon containing *I. turgida*(Kozur, Mostler and Rahimi-Yazd).According to the conodont fauna of Bed 29a, we suggest that the conodont fauna belongs to the *Isarcicella isarcica* Zone, due to the appearance of *Isarcicella turgida*.   This zone corresponds to the "mixed bed 3" of Sheng *et al.* [10,11], the "Boundary bed 3" of Wang [14], the top of the "Upper transitional bed" of Yin *et al.* [22, 24] and the lower part of the *Pseudoclaraia wangi-Ophiceras* Zone of Yang *et al.*[20]. *Isarcicella isarcica* (Huckriede) is widespread across the world. *Hindeodus parvus* (Kozur and Pjatakova) and *Isarcicella isarcica* (Huckriede) are worldwide species. They can be recognized from Asia, Europe and America.

CLARKINA LINEAGE AND *HINDEODUAS-ISARCICELLA* LINEAGE NEAR THE PERMIAN-TRIASSIC BOUNDARY AT THE MEISHAN SECTION

Evolution of the Clarkina Lineage

The *Clarkina* stock at section D of Meishan, Changxing appeared only in the Upper Permian and represented the descendant of the Eurasian stock [1]. It is mainly composed of *Clarkina orientalis* (Barskov and Koroleva), C. *subcarindata* (Sweet), C. *wangi* (Zhang), C. *changxingensis* (Wang and Wang), C. *deflecta* (Wang and Wang), C. *carinata* (Clark) and C. *dicerocarinata* (Wang and Wang). The distribution of the species of *Clarkina* in Changxingian at section D of Meishan is shown in Table 1. The evolution of the *Clarkina* stock is discussed as follows (Fig. 2):

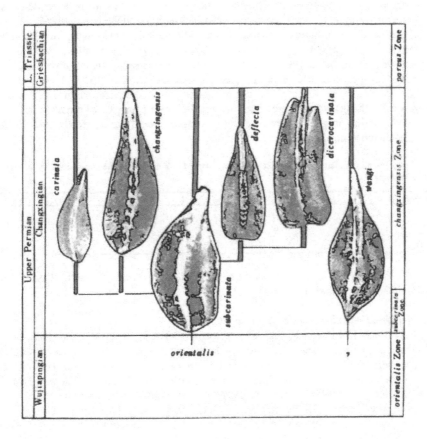

Figure 2. Evolution of the *Clarkina* lineage

Table 1. Distribution of species of *Clarkina* in Changxing Fm. at section D of Meishan

Formation	Upper Longtan Fm.	Changxing Fm.																								
Bed number	1	2	3	4	5	6	7	8	9	10	11	12	13	14	15	16	17	18	19	20	21	22	23	24	25	26
Clarkina carinata											4	1	7		4				3			17	3	3	1	
C. changxingensis						1			1		2	4	27						6		1	90	119	2	4	
C. deflecta											3	1	7	1	5							10	14	6	1	2
C. dicerocarinata															1							1	2			
C. wangi								8			9	13	10		1				2			29	9			
C. orientalis											1											1		2		
C. subcarinata		2	1	15		9		2	80		35	23	37		8			1				44	19			

Clarkina orientalis(Barskov and Koroleva) has been found in the Upper Longtan Formation at the Meishan section, Changxing. So, *Clarkina orientalis* (Barskov and Koroleva) is still the ancestor of C. *subcarinata* (Sweet). C. *subcarinata* (Sweet) diverged into three branches during the Changxingian. They are *Clarkina changxingensis* (Wang and Wang), C. *deflecta* (Wang and Wang) and C. *carinata* (Clark). The Pa element of C. *changxingensis* (Wang and Wang) is a symmetrical unit. The species ranges from the middle of the Lower Changxing Formation to the top of the Upper Changxing Formation. Pa elements of *Clarkina deflecta* (Wang and Wang) prominently have a symmetry transition (Table 2). The Pa elements in which the carina intersects the posterior margin at its center, at its left and at its right, are said to be approximate symmetrical, sinistral and dextral projection of the posterior margin. C. *deflecta* (Wang and Wang) ranged from the middle of the Lower Changxing Formation to the top of the Upper Changxing Formation. *Clarkina deflecta* (Wang and Wang) further developed into *Clarkina dicerocarinata* (Wang and Wang) by the divergence at the posterior end of the Pa element. *Clarkina carinata* (Clark) evolved from C. *subcarinata* (Sweet) by reduction of one side or two sides of the posterior margins of the platform. *Clarkina wangi* (Zhang),which ranged from the middle Lower Changxing Formation to the Upper Changxing Formation, might have evolved from *Clarkina leveni* (Kozur, Mostler and Pjatakova), but this postulation is unsettled.

Table 2 Posterior-end symmetry classification of *Clarkina deflecta* Pa element in the Changxing Fm. at section D of Meishan, Changxing

Bed number	6	11-2	13-1	15-1	15-2	22	23	24e	25	26	Total
Dextral	1				1	2	3	3	1	1	11
Sinistral		2			1	1	2	1			7
Approximate symmetrical		1	1	1		2	3			1	11

To sum up, the thriving stage of evolution of the   *Clarkina* stock was in the Upper Permian, especially in the Changxingian

Evolution of the Hindeodus-Isarcicella Lineage

Hindeodus parvus (Kozur and Pjatakova) is a species between *Isarcicella* and *Hindeodus*. The form of the *parvus* Pa element is very similar to the species of *Hindeodus* and no ramiform of *parvus* has been published that seems very like to belong to a species of *Isarcicella*. The attribution of H. *parvus* (Kozur and Pjatakova) is unsettled. In this paper, H. *parvus* (Kozur and Pjatakova) is attributed to *Hindeodus*. H. *parvus* (Kozur and Pjatakova) ranges from the basal Triassic, and is earlier than *Isarcicella isarcica*(Huckriede). Thus , H. *parvus* is very significant in the Permian-Triassic boundary strata and in the evolution of the *Hindeodus-Isarcicella* lineage. But now, *Hindeodus latidentatus* (Kozur, Mostler and Rahimi-Yazd) is found in Bed 25 (white clay) of the top of the Permian at section B of Meishan, Changxing. We consider that H. *latidentatus* is a   forerunner of the *Isarcicella* stock. The Pa element of H. *latidentatus* and the Pa element of *H. parvus* are very similar, except that the height of the denticles of *H. parvus* is higher than the denticles of *H .latidentatus*. Orchard[8] assigned *Hindeodus latidentatus* from Bed 25 in section B at Meishan to *H.* aff *H. parvus*. But, in our view *H. latidentatus* extended from the Late Changxingian to the Early Griesbachian, and *H. parvus* appeared only in the Griesbachian. The specimen from Bed 25 of section B at Meishan has a closer affinity to *H. latidentatus* than to *H. parvus*. According to the conodont data of the Meishan section, Changxing, the evolution of the *Hindeodus-Isarcicella* lineage is discussed as follows (Fig. 3):

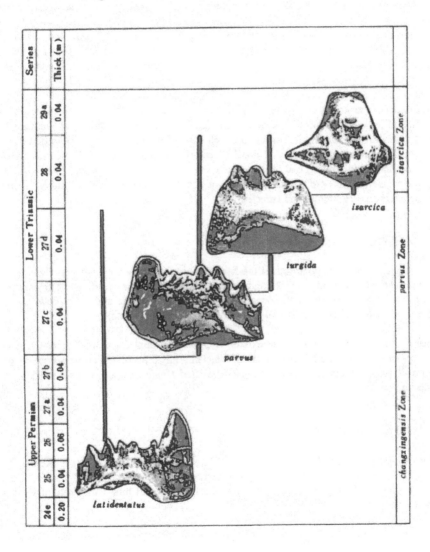

Figure 3. Evolution of the Pa element of *Hindeodus-Isarcicella* lineage at the Meishan

Isarcicella isarcica(Huckriede) from the bed 28 at the section A of Meishan, *I. turgida* (Kozur, Mostler and Rahimi-Yazd) occurred from the bed 882-3 of Zhongxin Dadui quarry section of Meishan (after Wang, 1994), *H. parvus* (Kozur *et* Pjatakova) from the bed 27c at section D of Meishan, and *H. latidentatus* (Kozur, Mostler and Rahimi-Yazd) from the bed 25 at the top of the Permian at the section B of Meishan.

The *Hindeodus-Isarcicella* lineage is mainly composed of *H. latidentus*, *H. parvus*, *Isarcicella turgida* and *I. isarcica*. *Hindeodus latidentatus* occurred from Bed 25 (white clay) near the top of the Upper Permian at section B of Meishan, and was contemporary with *Clarkina deflecta* and *C. changxingensis*. It can be confirmed that *H. latidentatus* occurred lower in the *H. parvus* zone or occurred in the upper Upper Permian. The Pa element of *H. latidentatus* discussed in this paper is distinguished from the Pa element of *H. parvus* by the ratio of length and width of unit and the length of the denticles. *Hindeodus latidentatus* developed into *H. parvus* by increasing the length of denticles, except the big denticle, and the width of the basal cavity was inflated. *H. parvus* occurred from Bed 27c, just at the Permo-Triassic boundary at section D of Meishan, Changxing. *H. parvus* occurred extensively at the base of the Lower Triassic and predated *I. isarcica*. *I. turgida* evolved from *H. parvus* by bearing a transverse ridge on the sides of its upper surface, and was found in the Bed 882-3 (equal to the Bed 27c at section D of Meishan) at the section at Zhongxin Dadui Quarry of Meishan [14]. According to the data of Central Iran [4] and NW Iran [5] the appearance of *I. turgida* is later than *I. isarcica*. But, at the Gartnerkofel-1 core section, Carinc Alps, Austria, *I. turgida* occurs earlier than *I. isarcica*. *I. turgida* evolved into *I. isarcica* by bearing a denticle on one side of its upper surface. *I. isarcica* is first found in Bed 28 at section A of Meishan, Changxing. Finally, *I. isarcica* bearing a denticle on each side of its upper surface, respectively, appears to represent the youngest member of the *Hindeodus-Isarcicella* stock. The *Hindeodus-Isarcicella* stock became extinct by the end of the Griesbachian.

As stated above, the evolutionary lineage of the *Hindeodus-Isarcicella* stock is composed of *latidentatus-parvus-turgida* and *isarcica*.

CONCLUSION

Based on abundant conodont data of the Meishan section, there are three conodont zones near the Permian-Triassic boundary, which can be correlated regionally or worldwide. They are in ascending order: *Clarkina changxingensis* Zone, *Hindeodus parvus* Zone and *Isarcicella isarcica* Zone. Among them, the *Clarkina changxingensis* Zone contains three faunas.

Meanwhile, the authors reveal the existence of the conodont *Clarkina* lingeage and *Hindeodus-Isarcicella* lineage at the Meishan section. It has not only proved the Permian-Triassic stratigraphic continuity at the Meishan section, but also testified *Hindeodus parvus* as a marker of the basal Triassic.

ACKNOWLEDGMENT

Thanks are due to Dr. Lucas, Curator of Paleontology and Geology, New Mexico Museum of Natural History and Science, for giving comments on this manuscript. We wish to thank Mrs. Tu Lijuang and Mr. Huang Jianyong, Faculty of Earth Sciences, China University of Geosciences, for typing this manuscript.

The paper is made possible through the support of the National Science Foundation of China.

REFERENCES

1. D. L. Clark and F. H. Behnken. Evolution and taxonomy of the North American Upper Permian *Neogondolella serrata* complex, *Journal Paleontology* 53:2, 263-275(1979).
2. Ding Meihua. Conodont sequences in the Upper Permian and Lower Triassic of South China and the nature of conodont faunal changes at the systemic boundary. In: *Permo-Triassic events in the eastern Tethys*. W. C. Sweet,

Yang Z . Y., J. M. Dickins and Yin H. F. (Eds.). pp. 109-119. Cambridge University Press(1992).

3. Ding Meihua, Zhang Kexin, Lai Xulong. Evolution of Clarkina Lineage and *Hindeodus-Isarccella* Lineage at Meishan Section, South China. In: *The Palaeozoic-Mesozoic Boundary Candidates of Global Stratotype Section and Point of the Permian-Triassic Boundary*. Yin H F (Ed.). pp. 65-71. China University of Geosciences Press(1996).

4. Iranian-Japanese Research Group. The Permian and the Lower Triassic Systems in Abadeh region, Central Iran, *Memoir of Faculty of Science, Kyoto University, Ser. Geol. Min.* 47:2,61-133(1981).

5. H. Kozur, H. Mostler and A. Rahimi-Yazd. Beitrage zur Mikrofauna permostriadischer Schichtfolgen. Teil II : Neue Conodonten aus dem Oberperm und der basalen Trias von Nord-und Zentraliran, *Geologie Palaontologie Mitteilung Innsbruck* 5:3,1-23(1975).

6. Lai Xulong, Ding Meihua and Zhang Kexin. The significance of the discovery of *Isarcicella isarcica* at the Meishan Permian-Triassic boundary stratotype Section in Changxing, Zhejiang Province. *Exploration of Geosciences* 11, 7-11, Press of China University of Geosciences, Wuhan(1995).(in Chinese with English abstract).

7. T. Matsuda. Early Triassic conodonts from Kashmir, India, Part 1: "*Hindeodus* and *Isarcicella*", *Journal of Geosciences, Osaka City University* 24:3,75-108(1981).

8. M. J. Orchard. Conodont fauna from the Permian-Triassic boundary: Observations and Reservations, *Permophiles, A Newsletter of SCPS* 28, 36-39(1996).

9. H. P. Schoenlaub. Permian-Triassic of the Gartnerkofel-1 Core (Carnic Alps, Austria): Conodont biostratigraphy, *Abh. Geol. Bundesanstalt* 45,79-98(1991).

10. Sheng Jinzhang, Chen Chuzhen, Wang Yigang, Rui Lin, Liao Zhuoting, Y. Bando K. Ishi, K. Nakamura and K. Nakamura. Permian-Triassic boundary in middle and eastern Tethys, *Journal Faculty Science, Hokkaido University, Ser.* 4,21:1,133-181(1984).

11. Sheng Jinzhang, Chen Chuzhen, Wang Yigang, Rui Lin, Liao Zhuoting, He Jinwen, Jiang Nayan and Wang Chenyuan. New advances on the Permian-Triassic boundary of Jiangsu, Zhejiang and Anhui. In: *Stratigraphy and Palaeontology of Systemic boundaries in China, Permian-Triassic Boundary*, Nanjing Institute of Geology and Palaeontology, Academia Sinica (Ed.).1, pp.1-22; Nanjing University Press(1987). (in Chinese with English abstract).

12. W. C. Sweet. Permian and Triassic conodonts from a section at Guryul Ravine, Vihi District, Kashmir, *Paleontological Contributions of University of Kansas* 49,1-10(1970a).

13. W. C. Sweet .Uppermost Permian and Lower Triassic conodonts of the Salt Range and Trans-Indus Ranges, West Pakistan. *University of Kansas, Department of Geology, Special Publications* 4,207-275(1970b).

14. Wang Chenyuan. A conodont-based high-resolution eventostratigraphy and biostratigraphy for the Permian-Triassic boundaries in South China, *Palaeoworld*, 4, 234-248. Nanjing University Press, Nanjing(1994). (in Chinese with English abstract).

15. Wang Chenyuan. Conodonts of Permian-Triassic boundary beds and biostratigraphic boundary, *Acta Palaeontologica Sinica* 34:2,129-151(1995). (in Chinese With English abstract).

16. Wang Chenyuan and Wang Zhihao. Permian conodonts from Longtan Formation and Changhsing Formation of Changxing, Zhejiang and their stratigraphical and palaeoecological significance. *Selected Papers on the 1st Convention of Micropalaeontological Society of China*. pp. 114-120. Science Press, Beijing (1979). (in Chinese with English abstract).

17. Wang Chengyuan, H. Kozur, Ishiga Hiroaki, G. V. Kotlyar, Ramovs A, Wang Zhihao, Y. Zakharov. Permian-Triassic boundary at Meishan of Changxing County, Zhejiang Province, China — A proposal on the global stratotype section and point (GSSP) for the base of Triassic. *Acta Micropalaeontologica Sinica* 13:2,109-124(1996).

18. Yang Shouren, Wang Xinping, Hao Weicheng. Permian conodonts from the "*Hypophiceras*" bed in Zhejiang, Jiangsu Province and its significance, *Chinese Science Bulletin*, 38:16,1493-1497(1993). (in Chinese).

19. Yang Zunyi, Yang Fengqing, Wu Shunbao. The ammonoid *Hypophiceras* fauna near the Permian-Triassic boundary at Meishan section and in South China: Stratigraphic significance. In: *The Palaeozoic-Mesozoic boundary Candidates of Global Stratotype Section and Point of the Permian-Triassic Boundary*. Yin Hongfu (Ed.).pp. 49-56. China University of Geosciences Press, Wuhan (1996).

20. Yang Zunyi, Yin Hongfu, Wu Shunbao, Yang Fengqing, Ding Meihua, Xu Guirong. Permian-Triassic boundary stratigraphy and Fauna of south China. *Geological Memoirs, Series* 2:6, Geological Publishing House, Beijing (1987) (in Chinese with English abstract).

21. Yin Hongfu, Yang Fengqing, Zhang Kexin and Yang Weiping. A proposal to biostratigraphic criterion of the Permian-Triassic boundary, *Memorie della Societa Geologica Italiana* 34:86,329-344(1988).

22. Yin Hongfu, Wu Shunbao. Transitional bed-the basal Triassic Unit of South China, *Earth Science* 10:163-174(1985).(in Chinese with English abstract).

23. Yin Hongfu, Ding Meihua, Zhang Kexin, Tong Jinnan, Yang Fengqing, Lai Xulong. Late Permian-Middle Triassic ecostratigraphy of Yangtze platform and its margins, Science Press, Beijing (1995). (in Chinese with English abstract).

24. Yin Hongfu, Wu Shunbao, Ding Meihua, Zhang Kexin, Tong Jinnan and Yang Fengqing. The Meishan Section — Candidate of the Global Stratotype Section and Point (GSSP) of the Permian-Triassic Boundary (PTB), *Albertiana* 14:15-31 (1994).

25. Zhang Kexin. The Permo-Triassic conodont fauna in Changxing area, Zhejiang Province and its stratigraphic significance, *Earth Science* — *Journal of China University of Geosciences* 12: 2,193-200 (1987). (in Chinese with

English abstract).

26. Zhang Kexin, Ding Meihua, Lai Xulong, Liu Jinhua. Conodont sequences of the Permian-Triassic boundary strata at Meishan section, South China. In: *The Palaeozoic-Mesozoic boundary Candidates of Global Stratotype Section and Point of the Permian-Triassic boundary.* Yin Hongfu (Ed.).pp. 57-64 China University of Geosciences Press, Wuhan (1996).

27. Zhang Kexin, Lai Xulong, Ding Meihua, Wu Shunbao and Liu Jinhua. Conodont Sequence and its global correlation of Permian-Triassic boundary in Meishan section, Changxing, Zhejiang Province, *Earth Science — Journal of China University of Geosciences* 20:6, 669-676 (1995).

28. Zhao Jinke, Sheng Jinzhang, Yao Zhaoqi, Liang Xiluo, Chen Chuzhen, Rui Lin and Liao Zhuoting. The Changhsingian and Permian-Triassic boundary of South China, *Bulletion of Nanjing Institute of Geology and Palaeontology, Academia, Sinica* 2:1-112(1981). (in Chinese with English abstract).

Proc. 30ᵗʰ Int'l. Geol. Congr., Vol. 11 , pp. 163-170
Wang Naiwen and J. Remane (Eds)
© VSP 1997

Late Paleozoic Deep-Water Facies In Guangxi, South China And Its Tectonic Implications

WU HAORUO, WANG ZHONGCHENG
Institute of Geology, Academia Sinica, Beijing 100029, China
KUANG GUODUN
Guangxi Bureau of Geology and Mineral Resources, Nanning 530023, China

Abstract

Guangxi region lies in the South China, facing the South China Sea on the south and jointing with Vietnam on the southwest. Traditionally it is regarded as an area where shallow shelves coexisted with deep basins on the platform background during Late Paleozoic. In recent years we found the depositional alternations of the cherts, siliceous limestones and thick piles of pillowed or massive basalts in the western Guangxi. Conodont fossils indicate the sequence ranging from Late Devonian to Late Permian in age. The obvious negative Ce anomaly and other geochemical characteristics of the cherts mean that they formed in deep-water pelagic environment with hydrothermal input. The geochemical compositions of the basalts also show an oceanic within-plate setting. Meanwhile, in the southern Guangxi from a continuous radiolarian chert sequence 11 radiolarian assemblages and 10 conodont zones from Late Devonian to Late Permian were recognized. The radiolarian assemblages can well-be correlated with those from western Yunnan and Japan. Those discoveries suggest that during the Late Paleozoic most part of Guangxi be possibly occupied by the oceanic environment. Some areas in Guangxi are covered with Late Paleozoic weakly deformed platform-type carbonate rocks. However, the lack of terrigeneous detritus and the presence of negative Ce anomaly from the intercalated siliceous beds indicate a pelagic setting for these carbonate rocks. They might be accumulated on the submerged highlands which emerged sometimes and formed an archipelago between the Yangtze and Indochina blacks.

Keywords: Deep-water facies, Guangxi, Paleo-Tethys, Late Paleozoic, Tectonic setting

INTRODUCTION

Guangxi region lies in the South China, facing the South China Sea (Fig. 1). It is usually regarded as a typical area of Caledonian movement and after that it joined with the Yangtze craton to become a part of the South China Black [5, 7, 9]. In the southwest Guangxi borders on the Vietnam. It is generally accepted that the South China and Indochina Blacks are separated by the Red River belt (or Song Da and Song Ma Sutures) in the northern Vietnam, although the collision age between them is still controversial [2, 3, 8, 9, 13]. However, studies of the Late Paleozoic deep-water sediments in Guangxi suggest the presence of Paleo-Tethyan oceanic seaway between the Yangtze block (not the South China Black) and the Indochina block in that time, and not a single suture separating the two blocks but a group of small terranes covered with the wide area in Guangxi, China and northern Vietnam.

REGIONAL GEOLOGICAL SETTING

Fig. 2 shows a geological outline of Guangxi region. The Yangtze block occupies its northeastern margin. The Qinzhou-Yunkai terrane and Damingshan terrane [19] lie in its southern part. Between the Yangtze block and Qinzhou-Yunkai terrane large area of Guangxi is covered with Late Paleozoic carbonate platform deposits. The Late Paleozoic deep-water deposits only crop out in small areas between them. The former mainly occur in Guilin area of northeast Guangxi, Jingxi area of west Guangxi, Liuzhou area of central Guangxi and large area of the Damingshan terrane. In addi-tion, some small isolated Late Paleozoic carbonate plat-forms also appear in northeast Guangxi. All these carbonate layers are weakly deformed with very gentle dips. Only latest Devonian and early Carboniferous basaltic intercalations appear in a few places in Jingxi area and Damingshan terrane, the thickness is less than 20 meters.

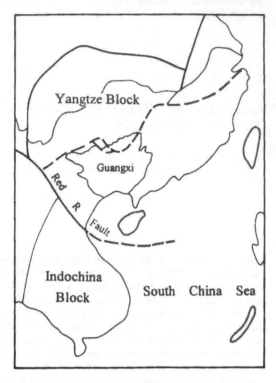

Figure 1. Location of Guangxi region.

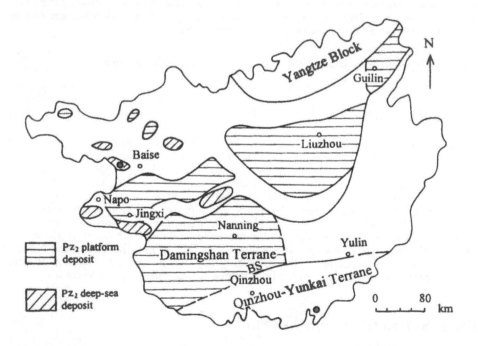

Figure 2. Sketch map showing geological features of Guangxi region. • Paleomagnetic sampling locality

On the other hand, the deep-water deposits in west Guangxi are strongly deformed and often associated with thick basaltic accumulations. In southern Guangxi, the Bancheng chert belt, from where 11 radiolarian assemblages and 10 conodont zones from Late Devonian were recognized, marks a suture between the Qinzhou-Yunkai terrane and Damingshan terrane [6, 19, 21]. Recently, cherts and pillow lavas were also reported from its eastern extension in Guangdong Province [23].

STRATIGRAPHY

The previous work on the Late Paleozoic deep-water deposits in Guangxi was limited to their sedimentological characteristics and facies analysis, the knowledge of their stratigraphy is incomplete. The microfossils is difficult to find and little was known about the microfossils in these strata. The lack of necessary biostratifgraphic data led to the consideration that the deep-water deposits were limited in Upper Devonian to Lower Carboniferous and Upper Permian, then the duration from Late Carboniferous to Early Permian was a stable stage in tectonic evolution [11]. Recently, we found a well exposed geological section along the Youjiang valley at the 30 km west of Baise (Fig. 3). It is consisted of Permo-Carboniferous alternations

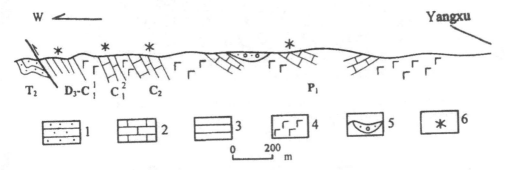

Figure 3. Late Paleozoic cross-section in Yangxu, west of Baise. 1. sandstone; 2. carbonate; 3. chert; 4. basalt; 5. Quaternary; 6. fossil locality.

of deep-water deposits and basalts. At the west end of the section, there are 60 m of gray to black bedded cherts, which is fault-contacted with the Triassic siltstones on the west. The earlier Early Carboniferous conodonts *Pseudopolygnathus triangularis* and *Siphonodella lobata* etc. were found from the cherts. They are followed by black and dark green basalts, amounting to 80 m in thickness. The basalts are overlain by the 26 m of thinly bedded dark gray chert and siliceous limestone, from which conodonts *Gnathodus honopuctatus*, *G. billineatus* and *G. typicus* etc. indicate later Early Carboniferous age. Therefore, the age of this basalt member is determined as Early Carboniferous. Upwards, the sequence is pass into siliceous limestones and fine-granied limestone turbidites (Fig.4), amounting to 40 m in thickness. The Late Carboniferous conodonts *Declinognathodus* was found from its top part. The Late Carboniferous to Early Permian strata are mainly consisted of alternations of carbonate rocks and basalts, but between the both always exist bedded cherts ranging in thickness from 1 to 10 m. The thickness of each basalt member from 6 m to more than 50 m. The carbonate deposites appear as dark-colored siliceous limestones (Fig. 5) or fine-grained distal turbidites with graded bedding. It has a total thickness of about 100 m. The Early Permian conodonts *Mesogondolella* was obtained from its upper part. Higher up the strata are not well exposed and dominated by mudstones and tuffites. Their age is supposed to be Late Permian based on the stratigraphical position.The similar Late Paleozoic deep-water sequence can be seen in the another section along the Youjiang valley at the 20 km west of Baise. Here the well-exposed

Upper Devonian strata are mainly consisted of dark-colored mudstones with siliceous shales and two basalt members with 100 m thickness each. The total thickness attains more than 500 m. Earlier Late Devonian conodonts *Mesotaxis* sp., *Hindeodella* sp., *Ozarkodina* sp., and *Ligonodina* sp. and later Late Devonian conodonts *Palmatolepis perlobata*, *P. arinutus* and *P. triangularis* were discovered from the lower and upper parts respectively.

Figure 4. Fine-grained limestone turbidite with graded bedding and wavy lamination, upper Lower Carboniferous, 30 km west of Baise, Guangxi.

Figure 5. Lower Permian siliceous limestone with chert intercalations, 30 km west of Baise, Guangxi.

The Late Paleozoic deep-water sediments are scattered at many places in western Guangxi. Almost of them are interbedded with basalts [18]. The lithology of these strata is similar to the above mentioned geological section at west of Baise, but their precise age determination require more detailed work.

The Late Paleozoic basalts are mainly distributed in the western Guangxi and include two types: massive basalt and pillow basalt. The former crop out in many localities and are conformably interbedded with the deep-water sediments, as the above mentioned sections in the Baise district. Their ages can be determined by the fossils obtained from the overlying and underlying sedimentary rocks. So far they occurred at least from Late Devonian to Early Permian. The pillow basalt covers a large areas in the Napo district (Fig. 6). Within the lavas there are many huge limestone blocks possibly slid from the seamount nearby. Meanwhile, some siliceous and pyroclastic turbiditic intercalations within basalts indicate the deep-water environment. The precise ages of the pillow basalts are difficult to ascertain due to their fault-contact with the surrounding Late Paleozoic deep-water sediments. However, the geochemical compositions are much alike that of the massive basalts and suggest an oceanic within-plate settings[16].

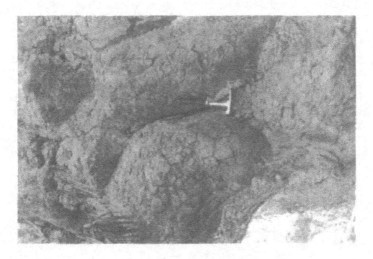

Figure 6. Well-preserved pillow structure of the Late Paleozoic basalt, 8 km of Napo County, Guangxi, China.

CERIUM ANOMALY AND PALEOMAGNETIC EVIDENCE

The geochemical compositions of the siliceous rocks are related to their depositional environment, of which the presence of negative Ce anomaly is an useful mark for the pelagic setting[10] . We already reported some geochemical data of the Late Paleozoic siliceous rocks in Guangxi [15, 22].The obvious negative Ce anomalies were obtained from the chert samples at many localities in western Guangxi .The Liuzhou district in central Guangxi is mainly covered with the Late Paleozoic carbonate platform deposit. However, the deep-water deposits also occur at some localities .At Tan Village, 14 km southwest of Liuzhou City, Upper Carboniferous consists of dark gray to black siliceous limestones with chert intercalations. In the northwest part of the Liuzhou City, along the Liujiang (Liu River) bank the middle-upper Permian strata appear as the alternations of dark gray to black chert and siliceous limestone with sedimentary structure of distal turbidite and contain volcanic clastics in the upper part. The negative Ce anomaly is also obvious for these siliceous rocks in the two localities. This fact and the lack of terrigeneous detritus in those platform type carbonates [14] suggest a pelagic setting for the Liuzhou area during the Late Paleozoic time. However, in the Bancheng belt of southern Guangxi only Upper Carboniferous-Lower Permian cherts show weak negative Ce anomalies and the Upper Devonian-Lower Carboniferous and later Permian cherts lack in negative

168

Ce anomaly[15]. This may implies that area was under the influence of the terrigeneous material from the Qinzhou-Yunkai terrane in the Late Paleozoic. On the other hand, this case supports the above mentioned negative Ce anomalies as the evidence of the pelagic setting for the western and central Guangxi during that time.

Moreover, the recent paleomagnetic results (Fig. 2) obtained by the joint research of Dr. X. Zhao in the University of California at Santa Cruz, US and us place the Qinzhou-Yunkai terrane at a paleolatitude of 20.6°S and the Baise area at 4.1°N during the Early Carboniferous time. This also implies a wide sea across thousands kilometers between the two places.

TECTONIC IMPLICATIONS

Above data indicate that the deep-water pelagic environment with extensive basaltic eruptions appeared in western Guangxi at least since Late Devonian and lasted for Carboniferous and Permian. The central Guangxi also in the pelagic setting and influenced by the volcanic activities at some times. In southern Guangxi the Upper Devonian-upper Permian siliceous sequence contains well-preserved radiolarian assemblages [21]. They can be well correlated with the contemporaneous radiolarian faunas in western Yunnan [20], northwestern Thailand [12] and Japan [4], of which the Permian fauna show the characteristics of the circumpacific deep-water radiolarian faunas [1]. Therefore, the most part of Guangxi was occupied by the deep-water pelagic environment and constituted an important seaway between the Yangtze and Indochina blocks (Fig. 7). The carbonate platforms in Liuzhou, Jingxi and Damingshan etc. were just the small submerged highlands, and didn't mean that Guangxi already developed into a stable platform in the Late Paleozoic.

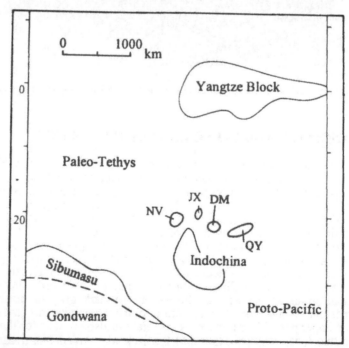

Figure 7. Palinspastic map of SE Asia in Late Paleozoic. DM: Damingshan Terrane; JX: Jingxi Terrane; NV: North Vietnam Terrane; QY: Qinzhou-Yunkai Terrane.

Moreover, the Devonian strata in western Guangxi rest directly on the Cambrian rocks and in a few case on the Lower Ordovician beds [17]. The stratigraphic sequences are similar to those in the adjacent Northern Vietnam block but different from those in the Yangtze block. These platforms or small terranes, including the Jingxi platform, the Liuzhou platform, the Northern Vietnam terrane, the Damingshan terrane and the Qinzhou-Yunkai terrane, together with the deep-water sediments between them constitute a wide and complicated transition area between the Yangtze and Indochina blocks.

ACKNOWLEDGMENT

Funding for this project was provided by the National Natural Science Foundation of China.

REFERENCE

1. R. Catalano, P. Di Stefano and H. Kozur. Permian circumpacific deep-water faunas from the western Tethys (Sicily, Italy) – new evidences for the position of the Permian Tethys, *Palaeogeog. Palaeoclimatol. Palaeoecol.* 87, 75-108 (1991).
2. Y. G. Gatinsky and C. S. Hutchison. Cathaysia, Gondwanaland and the Palaeotethys in the evolution of continental Southeast Asia, *Bull. Geol. Soc. Malaysia* 20, 179-199 (1987).
3. C. S Hutchison. Geological Evolution of Southeast Asia. Oxford Monographs on Geology and Geophysics 13, Clarendon Press, Oxford (1989).
4. H. Ishiga. Paleozoic radiolaria. In: *Pre-Cretaceous Terranes of Japan.* K. Ichikawa et al. (Eds.). pp. 285-296. Osaka City University (1990).
5. M. P. Klimetz. Speculations on the Mesozoic plate tectonic evolution of eastern China, *Tectonics* 2, 139-166 (1983).
6. Kuang Guodun, Li Jiaxiang, Zhong Keng, Su Yibao and Tao Yebin. New development on Carboniferous research in Guangxi, *Guangxi Geology*, 9, 17-32 (1996).
7. Liu Baojun, Xu Xiaosong, Pan Xingnun, Huang Huiqiong and Xu Qiang. Evolution and mineralization of ancient continental sedimentary crust in south china. Science Press, Beijing (1993).**
8. I. Metcalfe. Allochthonous terrane processes in Southeast Asia, *Phil. Trans. R. Soc. Lond. A* 331, 625-640 (1990).
9. I. Metcalfe. Pre-Cretaceous evolution of SE Asian terranes. In: *Tectonic evolution of Southeast Asia.* R. Hall and D. Blundell (Eds.). pp. 97-122. *Geol. Soc. Spec. Publ.* 106, London (1996).
10. R. W. Murray. Chemical criteria to identify the depositional environment of chert: general principles and applications, *Sedimentary Geology* 90, 213-232 (1994).
11. Ren Jishun. On the geotectonics of southern China, *Acta Geol. Sin.* 64, 275-288 (1990).*
12. K. Sashida, H. Igo, K. Hisada, N. Nakornsri and A. Apornmha. Occurrence of Paleozoic and Early Mesozoic radiolaria in Thailand (preliminary report), *Journal of Southeast Asian Earth Sciences* 8, 97-108 (1993).
13. A. M. C. Sengor, D. Altiner, A. Cin, T. Ustaomer and K. J. Hsu. Origin and assembly of the Tethyside orogenic collage at the expense of Gondwana Land. In: *Gondwana and Tethys.* M. G. Audley-Charles and A. Hallam (Eds.). pp. 119-181. *Geol. Soc. Spec. Publ.* 37, London (1988).
14. Sha Qingan, Wu Wangshi and Fu Jiamo (Eds.) Comprehensive research on Permian in Guizhou-Guangxi area and its petroleum potential. Science Press, Beijing (1990).
15. Wang Zhongcheng, Wu Haoruo and Kuang Guodun. Geochemistry and origin of Late Paleozoic chert in Guangxi and their explanation of tectonic environments, *Acta Petrol. Sin.* 11, 449-455 (1995).*
16. Wang Zhongcheng, Wu Haoruo and Kuang Guodun. Geochemical characteristics of the Late Paleozoic basalt from western Guangxi and its tectonic implication, *Acta Petrol. Sin.* (1996 in press).*
17. Wei Renyan. A new observation of the Cambrian stratigraphy in western Guangxi, *Geology of Guangxi* 2, 63-69 (1989).*
18. Wu Haoruo, Kuang Guodun and Wang Zhongcheng. Reinterpretation of basic igneous rocks in western Guangxi and its tectonic implications, *Sci. Geol. Sin.* 28, 288-289 (1993).
19. Wu Haoruo, Kuang Guodun, Xian Xiangyang, Li Yuejun and Wang Zhongcheng. The Late Paleozoic radiolarian chert in southern Guangxi and preliminary exploration on Paleo-Tethys in Guangxi, *Chinese Sci. Bull.* 39, 1025-1029 (1994).

170

20. Wu Haoruo and Li Hongsheng. Carboniferous and Permian radiolaria in the Menglian area, western Yunnan, *Acta Micropalaeontol. Sin.* **6**, 337-343 (1989).*

21. Wu Haoruo, Xian Xiangyang and Kuang Guodun. Late Paleozoic radiolarian assemblages of southern Guangxi and its geological significance, *Sci. Geol. Sin.* **29**, 339-345 (1994).*

22. Wu Haoruo, Kuang Guodun and Wang Zhongcheng. Late Paleozoic sedimentary tectonic settings in Guangxi, *Sci. Geol. Sin.* **32**, (1997, in press).*

23. Zhang Boyou, Shi Manquan, Yang Shufeng and Chen Hanlin. New evidence of the Paleo-Tethyan orogenic belt on the Guangdong-Guangxi border region, south china, *Geol. Rev.* **41**, 1-6 (1995). *

*   in Chinese with English abstract
** in Chinese

Proc. 30ª Int'l. Geol. Congr., Vol. 11 , pp. 171-180
Wang Naiwen and J. Remane (Eds)
© VSP 1997

Morphological change of Late Permian Radiolaria as seen in pelagic chert sequences

KUWAHARA KIYOKO

Department of Geosciences, Osaka City University, Osaka 558, Japan

Abstract

Morphological change of Late Permian *Albaillella* (Radiolaria) was studied in stratigraphic sections of bedded cherts of the Mino Belt, central Japan. These cherts are considered to have formed in a pelagic condition in the Panthalassa Ocean. Time duration of the studied sections is estimated less than several million years. Morphometric study was achieved in five species of *Albaillella* : *A. triangularis*, *A. excelsa*, *A. flexa*, *A. lauta* and *A. levis*. Measured parts are HU: height of upper part, HL: height of lower part, WA: width between rods at aperture, and AA: bending angle of apical part. Morphological change of *Albaillella* is interpreted to be affected by both evolutionary and environmental factors. Mean shell size is larger at the Ryozen section than the Gujo-hachiman section. It may show the differences in local environments between the two areas. Temporal changes both shell shape and size are detected in the sections. Gradual morphological changes in HL and AA are regarded as the evolutionary changes of *Albaillella*. The similar fluctuations in mean HU are commonly found between co-occurred species, reflecting the environmental change with several hundred thousand years duration within Late Permian time.

Keywords : Radiolaria, morphological change, Permian

INTRODUCTION

Well-preserved radiolarian assemblages are detected from Upper Permian bedded chert of the Mino Belt, in the Inner Zone of Southwest Japan. Stratigraphic sequence of bedded chert is suitable for studying radiolarian morphological change through time. Late Permian radiolarian assemblages in the Mino Belt are characterized by Albaillellaria, Entactinaria, Spumellaria, stauraxon polycystine, etc. Spherical radiolarians, such as Entactinaria and Spumellaria, are most popular in the assemblages, whereas Albaillellaria is not common. However, Albaillellaria is the most important group for radiolarian biostratigraphy in Permian. The genus *Albaillella* occurs most abundant in Albaillellaria and its stratigraphic distribution has been examined in detail [6, 8, 9]. There are changes in abundance and relative frequency up section. The highest frequency of occurrence changes from *A. triangularis* Ishiga, Kito and Imoto, *A. excelsa* Ishiga, Kito and Imoto, *A. flexa* Kuwahara, then to *A. levis* Ishiga, Kito and Imoto [8, 9]. The highest frequency part was interpreted to represent the prosperity period (acme) of species. The species of Late Permian *Albaillella* are regarded as a monophyletic group by their stratigraphic distribution and overall shell geometry. They have a simple shell and only one ventral wing, so it is easy to compare the shell geometry among species. Kuwahara [9] examined a change in morphology within the *A. excelsa-A. lauta* Lineage group. In this report, morphometrical study of Late Permian *Albaillella* is shown comprehensively. Temporal morphological change of *Albaillella* is interpreted to be affected by both evolutionary and environmental factors. The morphological change will give important information to reconstruct the phylogeny of *Albaillella* and paleoenvironments.

172

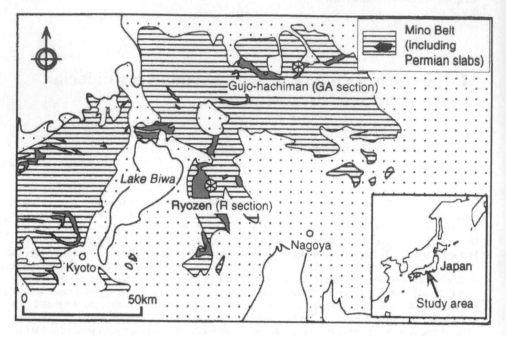

Figure 1. Index map of the studied sections. Geologic map of the Mino Belt is simplified from Wakita [17].

MATERIALS

Jurassic accretionary complex in the Mino Belt is widely distributed in central Japan (Fig. 1). This sedimentary complex was divided into six tectonostratigraphic units on the basis of composition, fabric and structure [17]. Studied materials were collected from the chert block of the Funafuseyama unit. The Funafuseyama unit is characterized by stacked slices of Middle Jurassic melanges and disrupted turbidites frequently associated with slices of greenstone, limestone and chert of Permian age. Two stratigraphic sections of bedded chert from the Gujo-hachiman and the Ryozen areas were selected. Fig. 2 shows columnar sections of the Gujo-hachiman (GA) and Ryozen (R) sections and the stratigraphic range of *Albaillella*.

Gujo-hachiman area (GA section)
The Gujo-hachiman area is located in the central part of the Mino Belt. Wakita [16] reported well-preserved Late Permian radiolarians from a chert block at locality G1709. This bedded chert is correlated to the *Neoalbaillella optima* and the *N. ornithoformis* Assemblage Zones [10]. *A. triangularis, A. excelsa, A. flexa,* and *A. levis* Abundance Zones were proposed in this section from lower to upper part [8]. Although thickness of this section is 9.7 meters, lower part with 6 meters thickness is treated herein. The same materials were used in my previous works [8-10]. Geological setting and materials were described therein. The section is named GA section (previously Gj section), and sixty-three samples were examined herein. Time duration of the GA section is estimated less than several million years [8].

Ryozen area (R section)
The Ryozen area is situated southwestern part of the Mino Belt. Materials were collected from the same locality as Ishiga et al. [6] that reported the *Neoalbaillella* Assemblage.

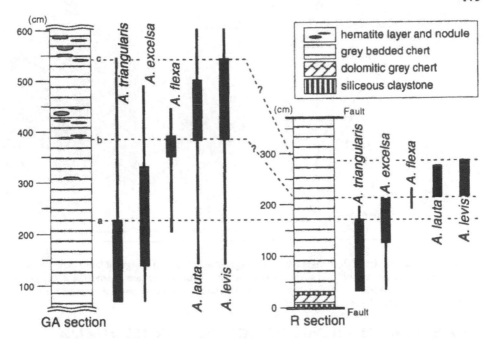

Figure 2. Columnar sections of the GA and R sections. Solid line shows the stratigraphic range of each species of *Albaillella*. Abundant occurrence is shown as a thick line. a: The top of the *A. triangularis* Abundance Zone; b: The base of the *A. levis* Abundance Zone; c: The top of the *A. levis* Abundance Zone.

Thickness of the section is 3.7 meters. The same materials with Kuwahara [8, 9] were reexamined. Forty-six samples were measured herein from the R section (Fig. 2). Time duration of the R section is also estimated less than several million years.

METHOD

Rock samples were treated in five percent hydrofluoric acid. Residues were collected by sieves of 35-200 mesh. Specimens of *Albaillella* were picked up on glass slides under binocular microscope, then glued with the "Entellan neu". Photocopies were taken for measurement under transmitted light microscope using video camera and video copy processing system. Five species of *Albaillella* (*A. triangularis*, *A. excelsa*, *A. flexa*, *A. lauta* and *A. levis*) were selected for this study. These species occur abundantly in the studied sections. Specimens of *Albaillella* were picked up bed by bed, except for the horizons of ill preservation. More than ten species, ideally more than twenty-five individuals were measured every stratigraphic horizon.

Under the transmitted light microscope, the internal rods of *Albaillella* are visible. The base line of measurement is a bisector line between the angle of ventral and dorsal rods (Fig. 3). Measured parts of shell are HU: height of upper part, HL: height of lower part, WA: width between rods at aperture, and AA: bending angle of apical part. HU is proximal part in the shell, and the morphological difference between species is comparatively small. The mean value and its ninety-five percent confidence intervals were computed.

174

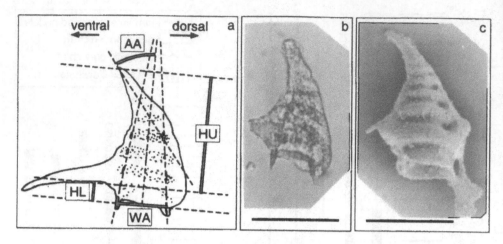

Figure 3. a. Measured parts of *A. triangularis*. HU: height of upper part; HL: height of lower part; WA: width between rods at aperture; AA: bending angle of apical part. b. Transmitted photomicrograph of *A. triangularis* from the GA section. Scale bar = 100μm. c. Scanning electronic photomicrograph of *A. triangularis* from the GA section. Scale bar = 100μm.

MORPHOLOGICAL CHANGE OF LATE PERMIAN *ALBAILLELLA*

Change of measured features through time
Graphs (Figs. 4-6) show temporal changes in measured features. Fig. 7 summarizes changes in mean HU of five species of Late Permian *Albaillella*. Studied sections are correlated with the *A. triangularis* and *A. levis* Abundance Zones [8].

HU (height of upper part): Mean value of HU of *A. triangularis* is 110μm in the GA section, changing from 100 to 118 μm gradually (Fig. 4). In the R section, mean HU of *A. triangularis* is 121μm and it ranges from 110 to 130 μm. HU of *A. excelsa* fluctuates around 110 μm in the GA section, measuring from 103 to 118 μm (Fig. 5). In the R section, there is also fluctuations in the HU of *A. excelsa*. HU of *A. flexa* fluctuates slightly in the GA section (Fig. 6). HU of *A. levis* fluctuates around 120 μm in the GA section, and it varies from 114 to 127μm. In the R section, the mean value is 130 μm and it ranges from 127 to 134 μm. HU of *A. lauta* shows a similar fluctuation pattern. In the GA section, it varies from 126 to 130 μm. In addition, HU seems to decrease within the range of species, as a general trend (Fig. 7).

HL (height of lower part): Mean of HL of *A. excelsa* tends to increase within lower part of the GA section. HL reaches 130μm in maximum, after that, it declines to 80-90 μm considerably (Fig. 5). A similar change is found at HL of *A. excelsa* from the R section.

WA (width between rods at aperture): In general, the values of WA remain stable.

AA (bending angle of apical part): Mean AA of *A. flexa* varies through time in GA section. The degree of AA is getting on for 60 degree in maximum (Fig. 6).

Figure 4.(→) Measurements of *A. triangularis* plotted against the thickness of the GA and R sections. Black squares show mean values from more than nine specimens for individual samples. White squares indicate mean values from less than ten specimens. Ninety-five percent confidence intervals for individual sample means are plotted as horizontal bars.

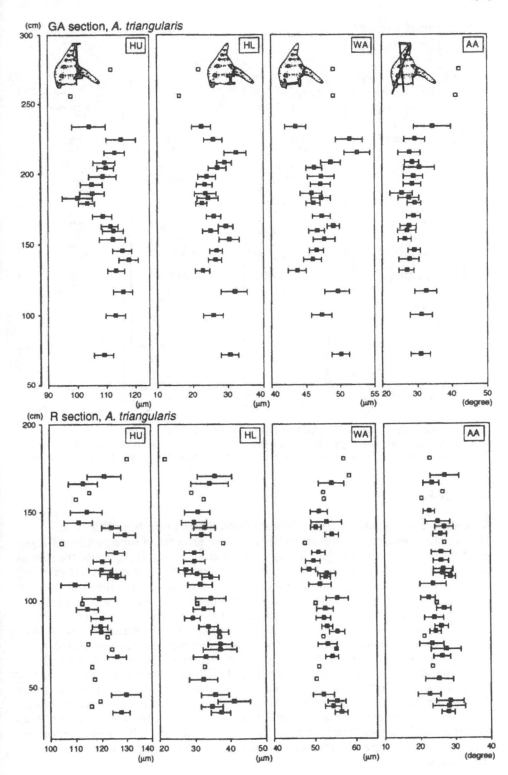

(cm) GA section, *A. triangularis*
(cm) R section, *A. triangularis*

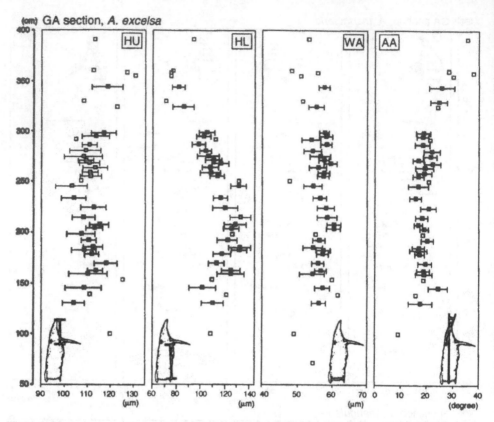

Figure 5. Measurements of *A. excelsa* plotted against the thickness of the GA section. Black squares denote mean values from more than nine specimens for individual samples. White squares show mean values from less than ten specimens. Horizontal bars indicate ninety-five percent confidence intervals for individual sample means.

DISCUSSION

Depositional environment of Upper Permian bedded chert in the Mino Belt

The Permian paleogeography is characterized by the presence of supercontinent Pangea, Panthalassa and Tethys Seas [e.g. 7, 13]. Barron and Fawcett [1] reviewed the climate simulations of the Permian, and the surface circulations of the oceans were also predicted as similar to the present-day Pacific. It was characterized the easterly currents at the lower latitudes. Under such situations, the bedded chert in the Mino Belt were formed successively in the pelagic condition in the Panthalassa. Paleomagnetic study of Triassic bedded chert at the Inuyama area shows that the chert sequences were formed in low latitudes [14]. The paleolatitude of Upper Permian chert also may indicate low latitudes. In Permian time, some seamounts were scattered in the ocean where bedded chert was deposited in the Mino belt. According to Sano [e.g. 12], limestone breccia as talus deposit is distributed around shallow marine sediments on the seamounts. The current system around the seamount might be complicated, producing large, regional differences in environment. There are some intercalations of siliceous claystone, dolomitic chert, and hematite nodules and lenses in the Upper Permian bedded chert in the Mino Belt. Their occurrences are variable in the stratigraphic horizons, suggesting regional differences in depositional environment.

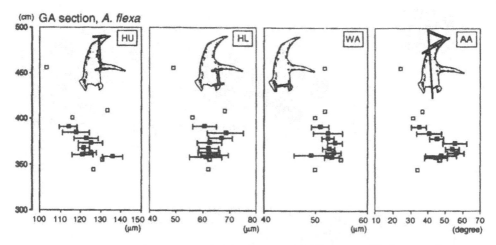

Figure 6. Measurements of *A. flexa* plotted against the thickness of the GA section. Black squares show mean values from more than nine specimens for individual samples. White squares denote mean values from less than ten specimens. Horizontal bars denote ninety-five percent confidence intervals for individual sample means.

The cause of morphological change of Radiolaria

The morphological changes of radiolarian shell may be influenced by following factors: (1) evolutionary (generic) change, (2) environmental change, (3)difference in fossilization, e.g. transportation, dissolution, deformation, etc. The factor (3) may be important, but it is not treated herein, because it is very difficult to evaluate in the bedded chert. Then, how can we distinguish the factor (2) from the factor (1)? The pattern of changes may give a hint.

T-test for equality of means

In order to compare the mean values of measured features between both sections, t-test was applied in ninety-five percent confidence intervals. The results are shown in Table 1. Eight means in size of the R section are larger than that of the GA section, on the other hand, only one mean of the GA section is larger than that of the R section. Mean shell size is larger at the Ryozen section than the Gujo-hachiman section, in general. The difference in overall shell size may be influenced by the paleoenvironmental conditions in each region.

Comparison of morphological change between co-occurred species

The morphological variation of upper part of shell is rather small in Late Permian *Albaillella*. HU contains proximal part of radiolarian shell and may be conservative. Fluctuation patterns in mean HU are similar between co-occurred species;*A. triangularis* - *A. excelsa* and *A. lauta* - *A. levis*, respectively (Fig. 7). Accordingly, they may reflect the paleoenviromnental change. Marked fluctuations in HU are found in rather short interval with several hundred thousand years within Late Permian time. On the other hand, the cause of the tendency to decrease in HU is not clear.

HL of *A. excelsa* and AA of *A. flexa* show long-term variations, associated with the change of overall shell geometry. These morphological changes are regarded as evolutionary trends.

178

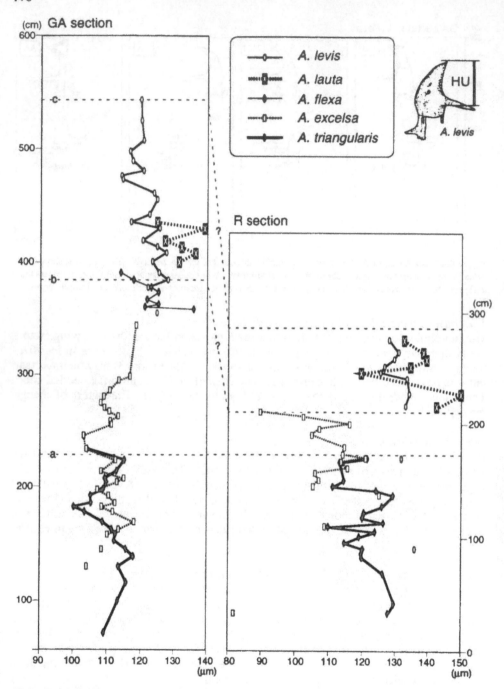

Figure 7. Mean HU of each species of Late Permian *Albaillella* plotted against the thickness of the GA and R sections. a: The top of the *A. triangularis* Abundance-Zone; b: The base of the *A. levis* Abundance-Zone; c: The top of the *A. levis* Abundance-Zone.

Table 1. Mean, standard deviation, and number of specimens of Late Permian *Albaillella* in the GA and R sections. T-test for equality of means was applied in ninety-five percent confidence intervals.

		GA section			R section			
		mean (μm)	S. D. (μm)	N	mean (μm)	S. D. (μm)	N	t-test
A. excelsa								
	HU	111.37	12.04	649	108.87	11.63	113	GA>R
	HL	114.01	24.18	803	127.45	30.62	114	GA<R
	WA	56.92	5.76	820	56.68	7.01	126	GA=R
	AA	19.20	7.52	730	20.55	6.37	131	GA=R
A. flexa								
	HU	122.78	12.66	199	121.72	14.05	5	GA=R
	HL	62.98	11.66	221	63.17	12.63	4	GA=R
	WA	52.41	4.10	215	52.17	2.28	5	GA=R
	AA	48.18	14.70	210	37.67	14.79	6	GA=R
A. lauta								
	HU	131.48	12.53	205	140.90	12.74	91	GA<R
	HL	69.28	11.55	257	69.94	12.06	93	GA=R
	WA	52.79	4.18	237	53.52	5.11	89	GA=R
	AA	27.10	7.79	217	26.76	7.15	92	GA=R
A. levis								
	HU	121.31	9.67	609	130.47	9.70	284	GA<R
	HL	31.00	7.88	757	32.39	6.00	302	GA<R
	WA	48.01	3.64	766	50.35	3.72	296	GA<R
	AA	32.20	6.69	669	32.73	7.44	290	GA=R
A. triangularis								
	HU	110.39	10.60	580	121.31	11.41	500	GA<R
	HL	26.44	7.80	736	32.88	8.50	649	GA<R
	WA	47.30	4.62	757	52.71	5.24	612	GA<R
	AA	28.93	8.06	633	25.76	6.89	666	GA>R

Granlund [2] studied the size and shape patterns in the Recent radiolarian *Antarctissa* from the South Indian Ocean transect. The shape variation of *Antarctissa* shows a close connection with the sea-surface temperature and salinity parameters. Granlund [3] examined the variations in size and shape of *Antarctissa* from piston cores in southeast Indian Ocean for the last 0.5 Ma. Oscillation in size and shape accords with oxygen isotope data. *Antarctissa* shows larger forms during glacial periods and smaller forms during interglacial periods. In short, the size pattern of *Antarctissa* indicates the paleoclimatological fluctuation.

In this study, the change in mean HU of Late Permian *Albaillella* is considered to show environmental change. Shell size of *Albaillella* may be also affected by water temperature. Further geochemical research will be necessary to specify a definite factor. The fluctuation pattern of mean HU may reflect the paleoenvironmental change with several hundred thousand years intervals. Origin of bedded chert has been discussed [e.g.4, 11]. Rhythmic bedded is considered to be genetically related to radiolarian blooming with the Milankovitch cycles [4, 15]. If so, radiolarian shell size also may be influenced by these fluctuations.

180

Acknowledgements

I wish to thank Prof. Dr. Yao A. for his advice during the course of this study. I wish to acknowledge Prof. Dr. Mizutani S. and Prof. Dr. Wu H. for giving me the opportunity to publish this paper. I wish to also thank Dr. Ezaki Y. for reviewing this manuscript.

REFERENCES

1. E.J. Barron and P.J. Fawcett. The climate of Pangaea: A review of climate model simulations of the Permian. In: *The Permian of Northern Pangea*. P.A. Scholle, T.M. Peryt and D.S. Ulmer-Scholle (Eds). pp. 37-52. Springer-Verlag, Berlin (1995).
2. A. Granlund. Size and shape patterns in the Recent radiolarian genus *Antarctissa* from a south Indian Ocean transect, *Mar. Micropaleontol.* 11, 243-250 (1986).
3. A. Granlund. Evolutionary trends of *Antarctissa* in the Quaternary using morphometric analysis, *Mar. Micropaleontol.* 15, 265-286 (1990).
4. Hori S.R., Cho C.-F. and Umeda H. Origin of cyclicity in Triassic-Jurassic radiolarian bedded cherts of the Mino accretionary complex from Japan, *The Island Arc* 3, 170-180 (1993).
5. Ishiga H. Late Carboniferous and Permian radiolarian biostratigraphy of Southwest Japan, *Jour. Geosci., Osaka City Univ.* 29, 89-100 (1986).
6. Ishiga H., Kito T. and Imoto N. Late Permian radiolarian assemblages in the Tamba district and an adjacent area, Southwest Japan, *Chikyu Kagaku* 36, 10-22 (1982).
7. G.D. Klein and B. Beauchamp. Introduction: Project Pangea and workshop recommendations. In: *Pangea: paleoclimate, tectonics, and sedimentation during accretion, zenith, and breakup of a supercontinent*. G.D. Klein (Ed.). pp. 1-12. Geol. Soc. America Spec. Paper 288 (1994).
8. Kuwahara K. Upper Permian radiolarian biostratigraphy -Abundance zone of *Albaillella* -, *News Osaka Micropaleontol.*, Spec. Vol., 10 (1996, in press). **
9. Kuwahara K. Evolutionary patterns of Late Permian *Albaillella* (Radiolaria) as seen in bedded chert sections in the Mino Belt, Japan, *Mar. Micropaleontol.* (1996, in press).
10. Kuwahara K. and Sakamoto M. Late Permian *Albaillella* (Radiolaria) from a bedded chert section in the Gujo-hachiman area of the Mino Belt, Central Japan. *Jour. Geosci., Osaka City Univ.* 35, 33-51 (1992).
11. E.F. Mcbride and R.L. Folk. Features and origin of Italian Jurassic radiolarites deposited on continental crust, *Jour. Sed. Petrology* 49, 837-868 (1979).
12. Sano H. Permian oceanic-rocks of Mino Terrane, Central Japan. Part IV. Supplements and concluding remarks, *Jour. Geol. Soc. Japan* 95, 595-602 (1989).
13. C.R. Scotese and R.P. Langford. Pangea and the Paleogeography of the Permian. In: *The Permian of Northern Pangea*. P.A. Scholle, T.M. Peryt and D.S. Ulmer-Scholle (Eds.). pp. 3-19. Springer-Verlag, Berlin (1995).
14. Shibuya H. and Sasajima S. Paleomagnetism of red cherts: A case study in the Inuyama area, central Japan, *Jour. Geophys. Res.* 91, 14105-14116 (1986).
15. Sugiyama K., Kamioka K. and Ozawa T. Depositional cyclicity and its origin in Triassic bedded chert, *Chikyu Monthly Extra* 10, 33-40 (1994). *
16. Wakita K. Allochthonous blocks and submarine slide deposits in the Jurassic formation southwest of Gujo-hachiman, Gifu Prefecture, central Japan, *Bull. Geol. Surv. Japan* 34, 329-342 (1983). **
17. Wakita K. Origin of chaotically mixed rock bodies in the Early Jurassic to Early Cretaceous sedimentary complex of the Mino terrane, central Japan, *Bull. Geol. Surv. Japan* 39, 675-757 (1988).

* In Japanese.
** In Japanese with English abstract.

Proc. 30ᵗʰ Int'l. Geol. Congr., Vol. 11, pp. 181-188
Wang Naiwen and J. Remane (Eds)
© VSP 1997

Clastic rocks in Triassic bedded chert of the Mino terrane, central Japan and the Samarka terrane, Sikhote-Alin, Russia

SATORU KOJIMA, KAZUHIRO SUGIYAMA,
Department of Earth and Planetary Sciences, Nagoya University, Nagoya 464-01, Japan
IGOR' V. KEMKIN, ALEXANDER I. KHANCHUK
Far Eastern Geological Institute, Russian Academy of Sciences, Vladivostok, Russia
and SHINJIRO MIZUTANI
Faculty of Social and Information Sciences, Nihon Fukushi University, Handa 475, Japan

Abstract

Clastic rocks rarely occur in Triassic bedded chert of the Mino terrane, central Japan and Samarka terrane, Sikhote-Alin, Russia, and are composed mainly of silt- to granule-sized fragments of basaltic (?) volcanic rocks, altered volcanic glass, chert, siliceous shale and radiolarian remains, with minor amounts of polycrystalline quartz, plagioclase, lutecite and glauconite (?). Conodont fossils are concentrated in some clastic rocks. Ages of the clastic rocks range from late Anisian to early Carnian on the basis of radiolarian and conodont fossils. The clastic materials were most probably derived from the basaltic volcanic edifices such as oceanic island/plateau and immature volcanic arc, and were transported by turbidity currents.

Keywords: Triassic, chert, Mino terrane, Samarka terrane, accretionary complex, radiolarian biostratigraphy, sedimentary event

INTRODUCTION

Upper Paleozoic to Mesozoic bedded chert formations in the circum-Pacific orogenic belt are generally considered to be free from coarse clastic materials. Middle to lower Upper Triassic bedded cherts, however, in the Mino terrane, central Japan and in the Samarka terrane, Sikhote-Alin, Far East Russia (Fig. 1) exceptionally include clastic-rock inter-layers and laminae [3, 5, 8, 11, 16]. Since the Triassic radiolarian cherts are considered to have been deposited in pelagic environments, such clastic rocks in the cherts must indicate a kind of particular sedimentary event in the Triassic ocean basins.

We report modes of occurrence, petrographic characteristics and radiolarian ages of these clastic rocks, and discuss the origin, provenance and geologic significance of the clastic rocks. Since the clastic rocks vary in grain size from siltstone to granule conglomerate, in this paper we collectively name the rocks as clastic rocks irrespective of their grain size.

GEOLOGIC SETTING

The clastic rocks occur in the Mino and Samarka terranes, both of which formed along the eastern continental margin of Asia by the accretionary processes in Jurassic time [7, 10, 12]. The Mino terrane is composed mainly of upper Paleozoic greenstone and limestone, Permian to Lower Jurassic bedded chert, and Jurassic clastic rocks and melanges with blocks of older ages. Rocks in the Mino terrane are subdivided into several units on the basis of the rock assemblage, age of accretion and deformation style [18]. The clastic rocks occur in the Kamiaso unit distributed in the southern part of the Mino terrane (Fig. 1). The Kamiaso unit consists mainly of Triassic to Lower Jurassic bedded chert and Middle to Upper Jurassic clastic rocks, and shows complicated geologic structure which is characterized by the repeated occurrence of chert, shale and sandstone formations by thrust faults [6, 19].

Figure 1. Index map showing the localities of cherts with clastic rocks, and also showing distribution of the Mino, Samarka and Nadanhada terranes (Jurassic accretionary complexes) and their metamorphosed equivalents.

The Samarka terrane shows lithology similar to that of the Mino terrane, and consists of Middle to Upper Jurassic turbidite and melanges with blocks of Middle Paleozoic ophiolite, Upper Paleozoic limestone, Upper Paleozoic and Lower Mesozoic bedded chert and basalt [4, 7, 13]. This terrane, together with the Nadanhada terrane in Northeast China (Fig. 1), was the northern extension of the Mino terrane before the opening of the Sea of Japan in Miocene time [7, 9].

OCCURRENCE AND PETROGRAPHY OF THE CLASTIC ROCKS

Ozaki Area in the Mino Terrane

The clastic rocks occur at several horizons in the Ozaki and Hisuikyo areas of the southern part of the Mino terrane. In the Ozaki area, the clastic rocks are 5 to 10 cm in thickness and show parallel lamination, cross lamination, graded bedding and small-scale channel structure [5]. The clasts are composed of sand- to granule-sized grains of basaltic (?) volcanic rocks, altered volcanic glass, chert, siliceous shale, polycrystalline quartz, and radiolarian and conodont remains. Very rarely occurs clastic plagioclase. Tsukamoto [16] reported the occurrence of a lutecite fragment from the clastic rock at Ozaki.

Hisuikyo Area in the Mino Terrane

In the Hisuikyo area, the clastic rocks occur as 0.5 to 2 cm thick clastic laminae in 5 to 10 cm thick chert beds, at least, at the 12 horizons in 15 m interval of the bedded chert [1, 2]. Clast composition is similar to that of the clastic rocks in the Ozaki area, although fragments of siliceous shale and chert are much more abundant than those of volcanic rocks. Sand-sized grains composed of glauconite (?) and/or magnesian siderite are often observed in this area. Clastic rocks themselves yield late Permian radiolarians like *Follicucullus scholasticus*, which are not embedded in the matrix, but are included in the chert and/or siliceous shale fragments [1].

Breevka Area in the Samarka Terrane

In the Breevka area about 100 km ENE of Vladivostok (Fig. 2), a clastic rock layer is embedded within the gray bedded chert formation (Figs. 3, 4A). Apparent thickness of the bedded chert is more than 30 m. Since melanges with blocks of chert and greenstone occur to the west of the exposure, this chert can be also regarded as a large block in the melanges. Part of the ophiolite (serpentinite, gabbro and amphibolite) can be seen about 5 km west of the chert exposure, although the relationship between the ophiolite and melanges is unclear.

The clastic rock layer is 10 to 15 cm in thickness, and shows clear graded bedding (Fig. 4A); grain size of the lower part is more than 0.5 mm (Fig. 4B), whereas that of the upper part is about 0.1-0.2 mm (Fig. 4C). The clastic grains are composed mainly of basaltic (?) rock fragments, altered volcanic glass, chert, siliceous shale, plagioclase, and conodont and radiolarian remains.

AGE OF THE CLASTIC ROCKS

We determined ages of the clastic rocks by extracting radiolarian fossils from cherts below and above the clastic rocks (Fig. 5). In order to determine and correlate the ages correctly, we use a detailed and newly developed Triassic radiolarian zonation [15]. In the Ozaki area radiolarian biostratigraphy of the cherts with clastic rocks was studied at three localities, and the ages are middle Ladinian (near the boundary between Zones 3B and 4A) at the locality Amaike-1, early Carnian (Zone 5A) at Amaike-2, and early Carnian (near the boundary between Zones 4B and 5A) at Ozaki. Clastic rock-bearing chert formations in the Hisuikyo area, range in age from late Anisian (Zone 3A) to early Ladinian (Zone 3B), and they are slightly older than the rocks in the Ozaki area. Bedded cherts in the Breevka area

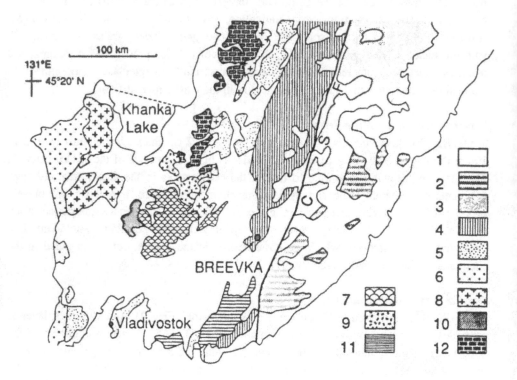

Figure 2. Geologic map of the southern part of Sikhote-Alin, Far East Russia. 1: Mesozoic sedimentary cover formations on Paleozoic terranes, Jurassic to Cretaceous volcanic rocks, and Cenozoic sedimentary cover formations; 2: Taukha terrane (Cretaceous accretionary complex), 3: Zhuravlevka terrane (Early Cretaceous turbidite terrane), 4: Samarka terrane, 5: Permian volcanic and sedimentary rocks, 6: Laoelin-Grodekov composite terrane (Permian island arc with Early Silurian volcanic and sedimentary rocks), 7: Devonian to Carboniferous plutonic, volcanic and sedimentary rocks, 8: Late Silurian volcanic and plutonic rocks, 9: Spassk terrane (Early Paleozoic accretionary complex), 10: Voznesenka terrane (Early Cambrian sedimentary rocks and Early Ordovician granite), 11: Sergeevka terrane (Proterozoic ? metamorphic rocks and Early Paleozoic plutonic rocks), 12: Matveevka-Nakhimovka terrane (Proterozoic metamorphic and plutonic rocks, and Early Paleozoic sedimentary rocks), CSF: Central Sikhote-Alin' Fault. For detailed geologic explanation, see [4, 13]. Modified from [4].

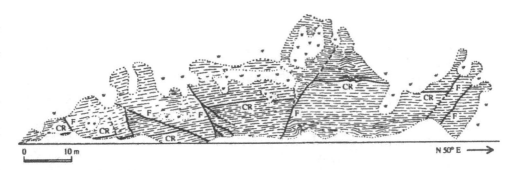

Figure 3. Sketch map showing occurrence of clastic rock in the Triassic bedded chert of the Breevka area, Sikhote-Alin. CR: clastic rock layer, F: fault.

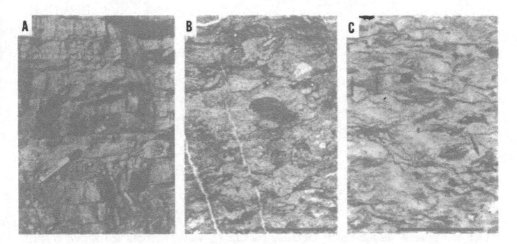

Figure 4. Photographs of the clastic rock in the Breevka area, Far East Russia. A: Field occurrence of the clastic rock (between the arrows). Note the lower coarser part with rough surface and the upper finer part with smooth surface. B: Photomicrograph of the lower part of the clastic rock, which is mostly composed of fragments of basaltic (?) rock and volcanic glass. Plane-polarized light. Scale bar is 1 mm. C: Photomicrograph of the upper part of the clastic rock. Arrows indicate conodont fossils. Plane-polarized light. Scale bar is 0.5 mm.

of Sikhote-Alin yield poorly-preserved radiolarian fossils, and are difficult to determine their ages correctly. Conodont fossils from the cherts below and above the clastic rock, however, indicate that the clastic rock is Ladinian, probably early Ladinian, in age [14].

DISCUSSION

Provenance of the Clastic Rocks

Since the clastic rocks embedded in the Triassic bedded chert of the Mino and Samarka terranes have sedimentary structures such as graded bedding, parallel lamination, cross lamination and channeling structure, the clastic grains must have been transported from the shallower provenance area by high energy water-mass flow like turbidity current. Most

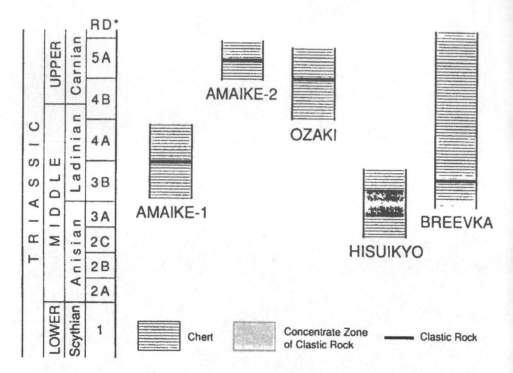

Figure 5. Correlation chart of the clastic rocks in the Mino and Samarka terranes. RD*: radiolarian zones after [15].

abundant fragments in the clastic rocks are basaltic (?) volcanic rocks and altered volcanic glass; basaltic (?) volcanic edifices, such as oceanic island/plateau and immature island arc, are most probable candidates for the source of the clastic materials. Chert and siliceous shale must have been deposited on and around the oceanic island/plateau or have been accreted along the island arc. Radiolarians from the chert and/or siliceous shale fragments in the Hisuikyo area indicate, at least, part of the siliceous rocks were deposited in late Permian time. The lutecite fragment found by Tsukamoto [16] provides climatic information of the provenance; evaporitic environment was prevailed or evaporites were distributed in the provenance. Glauconite (?) fragments from the Hisuikyo area might indicate shallow marine depositional environment. Although the majority of the clastic materials can be explained to be derived from the basaltic (?) volcanic edifices, polycrystalline quartz grains, 0.3 to 2.9 mm in length, found in the Ozaki area [16] imply continental source of the clasts.

Sedimentary Events Transporting the Clastic Materials into the Chert Depositional Basin
Since this type of clastic rocks are not common in Triassic bedded chert formations in the circum-Pacific orogenic belt, sedimentary events transporting the clastic materials into the chert depositional basins must be rare phenomena. Preliminary geochemical analyses [17] indicate that no Ir anomalies are detected on samples from the Ozaki and Hisuikyo areas. Neither shocked quartz nor impact spherules were found from the clastic rocks. These lines

of evidence exclude the extraterrestrial impact from the sedimentary events dispersing the clastic materials in the chert depositional basins. Volcanic activities are most probable events supplying the clastic materials, especially volcanic rocks and siliceous sediments deposited on and around the volcanoes. However, it is difficult to explain why the volcanoes were active only in late Anisian to early Carnian time; Permian to Early Jurassic bedded chert formations occur widely in the Mino and Samarka terranes. Moreover no clastic rocks have been found in upper Paleozoic to Mesozoic cherts of the circum-Pacific orogenic belt, although oceanic volcanism should be common phenomena in the Panthalassa ocean.

We found this type of sediments only in the late Anisian to early Carnian interval of the Permian to Early Jurassic chert. This means that the sedimentary events occur several times within 100 million years which is much longer than the history of human being. There must have been many events so far unknown to us. The sedimentary events supplying clastic materials into the chert basins might be one of such events that we don't know now.

Acknowledgements

We thank M. Adachi and M. Takeuchi for their critical review of early version of the manuscript, and S. Yogo for thin section preparation used in this study. Thanks are also due to V. S. Rudenko, A. N. Philippov, V. Golozubov, M. Kida, H. Ando, H. Tsukamoto, M. Kametaka, Y. Inoue, Y. Sakata and A. Ando for their discussion and help during the course of this study. Part of the field and laboratory works are financially supported by the Grant-in-Aid for Scientific Research of the Ministry of Education, Science and Culture of Japan (07238106, 08640567) and by the grant of the 29th International Geological Congress Financial Committee, given to S.K.

REFERENCES

1. H. Ando. Triassic bedded cherts including clastic grains in southern part of the Mino terrane, *Master Thesis of Nagoya Univ.* 14pp (1988).*
2. Y. Kakuwa. Lithology and petrography of Triasso-Jurassic bedded cherts of the Ashio, Mino and Tamba belts in Southwest Japan, *Sci. Pap., Coll. Arts and Sci., Univ. of Tokyo* **41**, 7-57 (1991).
3. Y. Kakuwa. Sedimentary petrographical, geochemical and sedimentological aspects of Triassic-Jurassic bedded cherts in Southwest Japan, *In A. Iijima, A.M. Abed and R.E. Garrison, eds., Proc. 29th Int. Geol. Cong., Part C*, 233-248 (1994).
4. A.I. Khanchuk, V.V. Ratkin, M.D. Ryazantseva, V.V. Golozubov, N.G. Gonokhova. Geology and mineral deposits of Primorye Kray (essay), *Dal'nauka, Vladivostok*, 66pp. (1996).
5. M. Kida. On the origin of bedded chert, with special reference to clastic beds in the Gifu-Kakamigahara area, central Japan, *Master Thesis of Nagoya Univ.* 36pp (1979).
6. S. Kido. Occurrence of Triassic chert and Jurassic siliceous shale at Kamiaso, Gifu Prefecture, central Japan, *News Osaka Micropaleont., Spec. Vol.* no.5, 135-151 (1982).*
7. S. Kojima. Mesozoic terrane accretion in Northeast China, Sikhote-Alin and Japan regions,

Palaeogeogr., Palaeoclimatol., Palaeoecol. 69, 213-232 (1989).

8. S. Kojima, Y. Inoue, K. Sugiyama and S. Mizutani. Clastic rocks in the Middle Triassic bedded chert in central Japan and Sikhote-Alin. *Jour. Geol. (Geol. Surv. Vietnam), Ser. B*, no.5-6, 158-159 (1995).

9. S. Kojima, K. Wakita, Y. Okamura, B.A. Natal'in, S.V. Zyabrev, Q.L. Zhang and J.A. Shao. Mesozoic radiolarians from the Khabarovsk complex, eastern USSR: their significance in relation to the Mino terrane, central Japan, *Jour. Geol. Soc. Japan* 97, 549-551 (1991).

10. S. Mizutani. Mesozoic evolution of the Japanese Islands. *Proc. 15th Int. Symp. Kyungpook National Univ.* 11-41 (1995).

11. S. Mizutani and Y. Koido. Geology of the Kanayama district with Geological Sheet Map at 1:50,000. Geological Survey of Japan, 111pp. (1992).*

12. S. Mizutani and S. Kojima. Mesozoic radiolarian biostratigraphy of Japan and collage tectonics along the eastern continental margin of Asia, *Palaeogeogr., Palaeoclimatol., Palaeoecol.* 96, 3-22 (1992).

13. B. Natal'in. History and modes of Mesozoic accretion in Southeastern Russia. *The Island Arc* 2, 15-34 (1993).

14. V.S. Rudenko. *personal communication* (1995).

15. K. Sugiyama. Triassic radiolarian biostratigraphy in the southeastern part of the Mino terrane, *Abst. 1995 Annual Meet. Palaeont. Soc. Japan* 124 (1995).**

16. H. Tsukamoto. Lutecite in Triassic bedded chert from the southern Mino terrane, central Japan, *Jour. Earth Sci., Nagoya Univ.* 36, 1-14 (1989).

17. T. Uchino, M. Ebihara, S. Kojima and S. Kawakami. Search for impact ejecta rocks in deep-sea pelagic sediments. *Abst. 1995 Japan Earth Planet. Sci. Joint Meet.* 462 (1995).**

18. K. Wakita. Origin of chaotically mixed rock bodies in the Early Jurassic to Early Cretaceous sedimentary complex of the Mino terrane, central Japan, *Bull. Geol. Survey Japan* 39, 675-757 (1988).

19. A. Yao, T. Matsuda and Y. Isozaki. Triassic and Jurassic radiolarians from the Inuyama area, central Japan, *Jour. Geosci., Osaka City Univ.* 23, 135-154 (1980).

* in Japanese with English abstract
** in Japanese

Proc. 30*ʰ* Int'l. Geol. Congr., Vol. 11 , pp. 189-199
Wang Naiwen and J. Remane (Eds)
© VSP 1997

Holarctic Fossil Mammals and Paleogene Series Boundaries

SPENCER G. LUCAS

New Mexico Museum of Natural History, 1801 Mountain Road N. W., Albuquerque, New Mexico 87104, U.S.A.

Abstract

Mammalian paleontologists have long attempted to correlate major turnover events in the evolution of Holarctic mammals to Paleogene Series boundaries. However, current understanding of the evolution and biostratigraphy of Holarctic Paleogene mammals indicates that no Paleogene Series boundary correlates to a major evolutionary turnover in mammals.

Keywords: Paleogene, fossil mammals, Europe, North America, Asia, Series

INTRODUCTION

Fossil mammals play a significant role in the correlation of nonmarine Paleogene strata. This is particularly true on the northern continents--North America, Europe and Asia--which formed a single, boreally-connected Holarctic landmass during much of the Paleogene.

Correlation by mammals relies heavily on the provincial biochronologies established as land-mammal "ages" (LMA's) (Fig. 1). First named in North America [43], LMA's are intervals of time that correspond to local faunas (assemblage zones) [14, 29]. Boundaries between LMA's usually correspond to turnover events among mammals due to evolutionary origination, extinction and/or immigration [44].

For many years, some mammalian paleontologists have argued that the Paleogene Series boundaries--especially the Paleocene/Eocene and Eocene/Oligocene boundaries--coincide with major turnover events in mammalian evolution [11, 15-16, 18]. Or, conversely, some have argued that major turnover events in mammalian evolution should be used to define the Paleogene Series boundaries. Here, I argue against these viewpoints by showing that no Paleogene Series boundary coincides with a major turnover in Holarctic mammalian evolution.

UPPER CRETACEOUS/PALEOCENE SERIES BOUNDARY

The Holarctic fossil record of mammals across the Upper Cretaceous-Paleogene Series boundary is very incomplete in Europe and Asia [7]. Only in North America is it dense enough to be of biostratigraphic value. Prior to the 1980's, this boundary was correlated to the boundary of the Lancian and

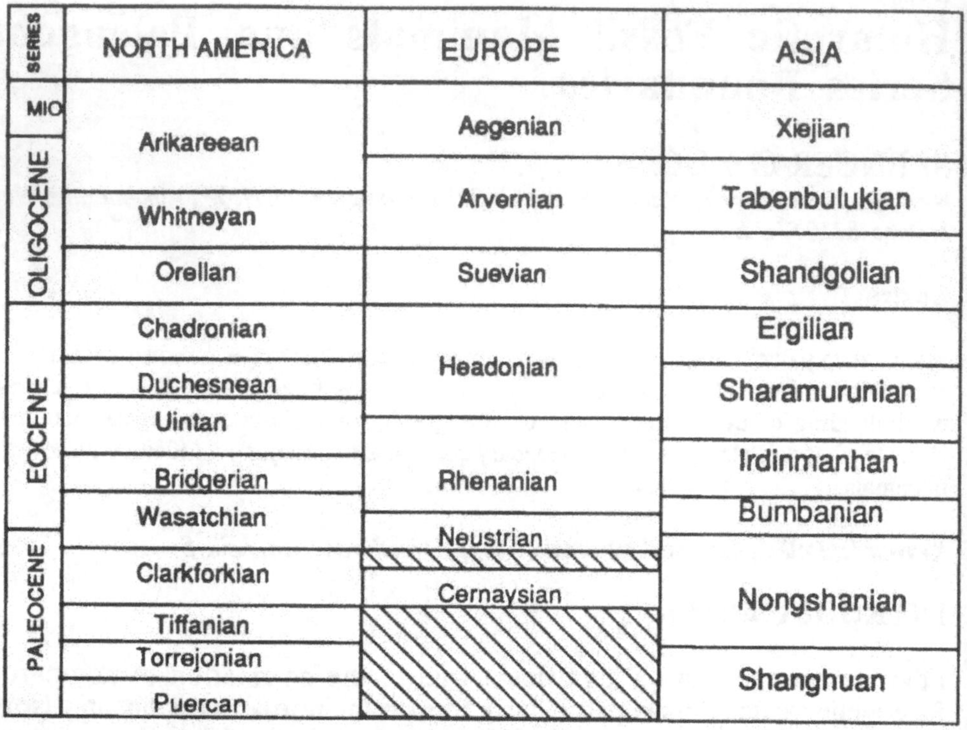

SERIES		NORTH AMERICA	EUROPE	ASIA
MIO		Arikareean	Aegenian	Xiejian
OLIGOCENE			Arvernian	Tabenbulukian
		Whitneyan		
		Orellan	Suevian	Shandgolian
EOCENE		Chadronian	Headonian	Ergilian
		Duchesnean		Sharamurunian
		Uintan		
		Bridgerian	Rhenanian	Irdinmanhan
		Wasatchian		Bumbanian
PALEOCENE			Neustrian	
		Clarkforkian	Cernaysian	Nongshanian
		Tiffanian		
		Torrejonian		Shanghuan
		Puercan		

Figure 1. Correlation of Paleogene LMA's of North America, Europe and Asia.

Puercan LMA's [30-31]. That boundary coincided with a perceived significant evolutionary turnover in mammals largely because known Lancian mammals actually predated the end of the Cretaceous and the classic Puercan mammals postdated the Cretaceous/Paleocene boundary, leaving a gap in mammal distribution of up to 2 Ma [21].

Detailed collecting across the Cretaceous-Paleogene boundary in northeastern Montana during the late 1970's and 1980's closed that gap, and a more refined biostratigraphic zonation of four LMA's across the Cretaceous-Paleogene boundary was proposed (Fig. 2). However, the two new LMA's-Bugcreekian [1] and Mantuan [39]-- were later judged to be not biostratigraphically distinct enough to warrant LMA status, so they were relegated to the status of zones or sub-land-mammal "ages" (SLMA).

Archibald and Lofgren [2] and Lofgren [19] most recently analyzed the record of mammals across the Upper Cretaceous/Paleocene Series boundary in northeastern Montana. Of the 28 mammalian genera present, 14 pass from Lancian to Puercan, only 4 have their LO (lowest occurrence) at the Puercan base and 10 have their LO within the Puercan. The base of the Puercan apparently is not a major turnover in mammalian evolution. A much more dramatic turnover corresponds to the younger, Puercan-Torrejonian boundary, which is of early Paleocene age [42]. Mammals thus do not

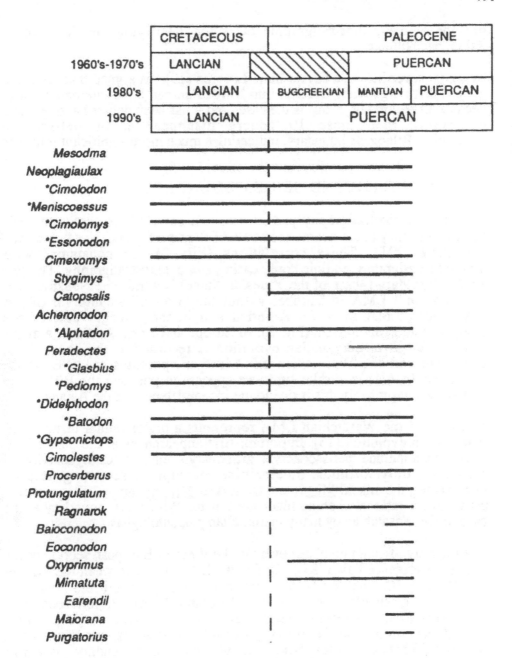

Figure 2. Stratigraphic ranges of mammalian genera across the Upper Cretaceous/PaleoceneSeries boundary in northeastern Montana [19]. Asterisks (*) mark genera whose Puercan occurrences may be due to reworking from Upper Cretaceous strata.

provide a strong biostratigraphic basis for identifying the Cretaceous-Paleogene boundary.

There is, nevertheless, a caveat here. Lofgren [19] has argued that 10 of the genera that range from Lancian into lower Puercan have Puercan records because of reworking. If this is the case (and it has been debated and remains uncertain), then these 10 genera disappear at or before the Cretaceous/Paleogene boundary, indicating a much more significant turnover event.

PALEOCENE/EOCENE SERIES BOUNDARY

Mammalian biostratigraphy provides robust correlations across the P/E Series boundary in the nonmarine strata of Western Europe, North America and Asia [22]. These correlations (Fig. 3) are consistent with magnetostratigraphy, radioisotopic dating and chemostratigraphy. The P/E Series boundary (=base of the Ypresian Stage) is within the Neustrian (= "Sparnacian") LMA in Europe, within the Graybullian interval of the Wasatchian LMA in North America and in the younger part of the Nongshanian LMA in Asia (Fig. 3). In Europe and North America, a major turnover event in mammalian evolution is recorded at the base of the Neustrian and the Wasatchian LMA's that predates the P/E Series boundary. In Asia, a similar turnover appears to postdate the P/E Series boundary, but probably this is due to the incompleteness of the Asian record.

The base of the Wasatchian LMA represents a major faunal turnover in mammalian evolution [15]. Important first appearances at the base of the Wasatchian include perissodactyls, artiodactyls, euprimates, hyaenodontid creodonts, didymoconids, didelphinine marsupials, several genera of insectivores, the miacid Miacis and the rodent Microparamys. This turnover provides the rationale behind placement of the P/E Series boundary at the base of the Wasatchian by many mammalian paleontologists.

The most significant faunal turnovers in the Western European section occur between Reference Level (RL) MP 6 and RL MP 7 and between RL MP 8/9 and RL MP 10. The two faunal turnovers readily delimit the Neustrian LMA. At the boundary between RL MP 6 and RL MP 7, perissodactyls, artiodactyls, chiropterans, euprimates, rodents and coryphodontid pantodonts appear, and the diversity of multituberculates, condylarths and plesiadapiforms decreases. Note, though, that this "boundary" is a gap between the Cernay (RL MP 6) fauna, which is a single point in Thanetian time, and the RL MP 7 assemblages, and the gap certainly accentuates the apparent evolutionary turnover (compare the North American record). At the boundary between RL MP 8-9 and RL MP 10, pantodonts and multituberculates disappear, and a profusion of lophiodonts dominates the fossil assemblages. In Asia, the base of the Bumbanian LMA corresponds to the appearance of perissodactyls, omomyid primates, rodents and *Hyopsodus* [9, 37].

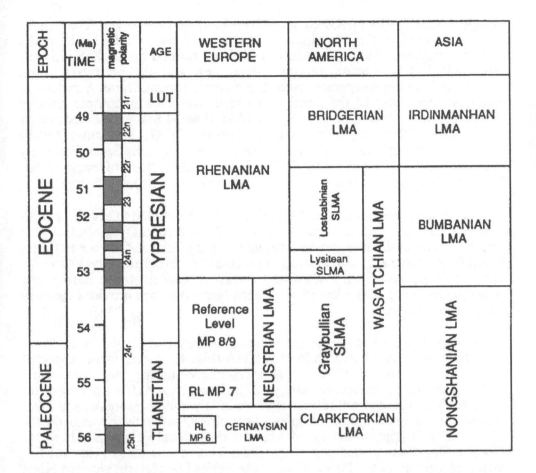

Figure 3. Correlation of Holarctic LMA's and SLMA's across the P/E Series boundary [22].

By definition, the base of the Eocene is the base of the Ypresian Stage. However, a GSSP (global stratotype section and point) has not yet been established for the P/E Series boundary. That GSSP may be placed so as to coincide with the base of the Ypresian Stage, or it may correspond to the nannoplankton NP 9/NP 10 Zonal boundary, the planktonic foraminiferal P5/P6 Zonal boundary or the carbon isotope excursion. These are the favored GSSP candidates [3], and all of them are younger than the mammalian turnover event. When a GSSP is established for the P/E Series boundary, that boundary will remain within the Neustrian, Wasatchian and Nongshanian LMA's.

EOCENE/OLIGOCENE SERIES BOUNDARY

In the 1980s, in accordance with the guidelines and statutes of the International Commission of Stratigraphy (ICS) of the International Union of Geological Sciences [8], International Geological Correlation Project 174 sought a suitable Eocene/Oligocene Series boundary stratotype. They chose the 19-m level in the Massignano section of Italy as the GSSP of the Eocene/Oligocene Series boundary, a decision ratified by the ICS [25-26]. The GSSP is in marine pelagic limestones, and the boundary point corresponds to the disappearance of the planktic foraminiferan *Hantkenina* and other members of the family Hantkeninidae. Radioisotopic ages of ashes just below this point are slightly older than 34 Ma, and the boundary point falls within magnetic polarity reversal C13r1. The most widely accepted earlier estimate of the age of the Eocene/Oligocene Series boundary was 36+ Ma [5], so the new GSSP has had the effect of moving the boundary so that it is more than 2 Ma younger than previously thought.

No direct biostratigraphic cross correlation has been made between fossil-mammal-bearing strata and marine strata with hantkeninid foraminiferans or other marine fossils that can be used to place the Eocene/Oligocene Series boundary. Instead, correlating mammalian biochronology to the Eocene/Oligocene Series boundary relies on radioisotopic dating and magnetochronology. The record in western North America provides the most precise correlation.

Previously, the Eocene/Oligocene Series boundary was placed in the lower part (or at the base of) the Chadronian LMA (Fig. 4), a placement supported by K-Ar ages from tuffs at Flagstaff Rim, Wyoming, and apparently consistent with magnetochronology [27]. Now, new Ar/Ar ages from the same tuffs as well as from tuffs elsewhere in Wyoming, Nebraska and West Texas, and a recalibrated magnetochronology, indicate the end of the Chadronian is approximately 34 Ma and thus closely corresponds to the Eocene/Oligocene Series boundary [28]. This means that in the last few years, placement of the Eocene/Oligocene Series boundary in western North America has shifted by more than 2 Ma and two magnetochrons (Fig. 4).

The most precise method now available to correlate the Eocene/Oligocene Series boundary in continental Asian strata is to correlate the Chadronian-Orellan LMA boundary to the land-mammal record in Mongolia and China. For many years, Russian and Mongolian vertebrate paleontologists considered the mammalian fossil assemblage of the Ergilin-Dzo svita in Mongolia to be characteristic of the early Oligocene [45]. They based this conclusion in part on the well-accepted correlation of the Ergilin-Dzo mammals with those of Chadronian age.

However, the Chadronian mammals are now considered late Eocene, as should correlative mammals in Asia, those of the Ergilin-Dzo svita. The oldest Oligocene mammal assemblage in Asia is well represented by the assemblage of the lower part of the Hsanda Gol svita, which is the mammalian fossil assemblage that temporally follows the Ergilin-Dzo mammals in Mongolia. This repositioning of the Eocene/Oligocene Series boundary has not been accepted by Chinese, Mongolian and Russian vertebrate paleontologists [11, 32, 41], but is well accepted in the West [4,

12-13] and followed here.

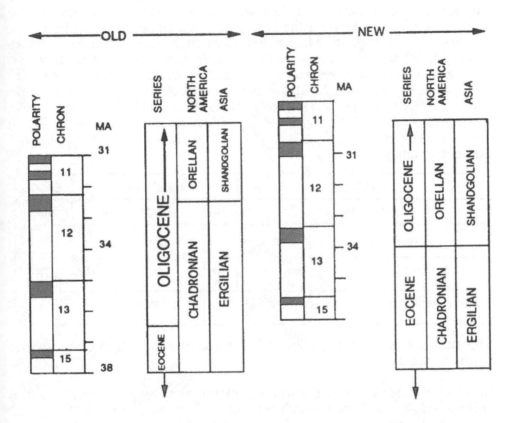

Figure 4. Old (pre-1992) and new (post-1992) correlation of North American and Asian LMA's across the Eocene/Oligocene Series boundary.

In North America and Asia, the Eocene/Oligocene Series boundary does correspond to an evolutionary turnover event in mammals reflected by its correspondence to a LMA boundary. However, this turnover is part of a protracted change in the mammalian fauna that began in the late Eocene and continued through the mid-Oligocene [4, 11-12, 40]. Similar, long-term changes also took place in the European mammalian fauna across the Eocene/Oligocene Series boundary [17]. Indeed, what Stehlin [33] called "la Grande Coupure" (the great break)--a supposed dramatic change in European terrestrial mammals across the Eocene/Oligocene Series boundary--actually is largely an artifact of dispersal lag [44].

OLIGOCENE/MIOCENE SERIES BOUNDARY

To my knowledge, there has not been a serious effort to equate the Oligocene/Miocene Series boundary with a mammalian turnover event. The base of the Miocene Series is the base of the Aquitanian Stage. In Europe, a succession of "Neogene" Mammalian Reference Levels labelled MN1 to MN17 [6, 24] actually crosses the Oligocene/Miocene Series boundary--RL MN1 and at least the lower part of RL MN2 are of late Oligocene age (Fig. 5). The oldest mammal localities that can be directly correlated to the Aquitanian stratotype belong to RL MN 2 [34]. This means the Oligocene/Miocene Series boundary is within the Aegenian LMA (Fig. 5) and does not correspond to a significant evolutionary turnover in the European mammalian fauna.

OLIGOCENE				MIOCENE			EPOCH	
CHATTIAN			AQ	BURDIGALIAN			STAGE	
ARVERNIAN	AGENIAN			ORLEANIAN			LMA	EUROPE
MP 25-MP30	MN1	MN 2a	MN 2b	MN3	MN4	MN5	MAMMAL LEVELS	
TABENBULUKIAN	XIEJIAN			SHANWANGIAN			LMA	ASIA

Figure 5. Correlation of Holarctic LMA's across the Oligocene/Miocene Series boundary.

In North America, correlation of mammalian assemblages to the Oligocene/Miocene Series boundary is achieved by magnetostratigraphy and radioisotopic dating. The boundary is correlated to within the Arikareean LMA, close to a turnover event that terminates the White River chronofauna and bracketed by minor immigration events [35-36].Thus, in North America, no major evolutionary turnover in mammals correlates to the Oligocene/Miocene Series boundary.

In Asia, the Oligocene/Miocene Series boundary has been placed to coincide with the boundary of the Tabenbulukian and Xiejian LMA's [20, 29, 38], However, this is a miscorrelation based on the misconception that the base of RL MN 1 correlates with the base of the Aquitanian. If correlation of the lower Xiejian LMA to RL MN 1 is correct [20, 23], then the Oligocene/Miocene Series correlates to a point within the Xiejian LMA and does not correspond to a significant evolutionary turnover in Asian mammals.

Acknowledgments
The National Geographic Society has funded my research on Asian Paleogene mammals since 1992. M.-P. Aubry, W. A. Berggren, H. de

Bruijn and R. J. Emry influenced many of the ideas presented here.

REFERENCES

1. Archibald, J. D. Latest Cretaceous and early Tertiary mammalian biochronology/biostratigraphy in the Western Interior, *Geol. Soc. Amer. Abs. Prog.* 19: 258 (1987).
2. Archibald, J. D. and Lofgren, D. L. Mammalian zonation near the Cretaceous-Tertiary boundary, *Geol. Soc. Amer. Spec. Pap.* 243: 31-50 (1990).
3. Berggren, W. A. and M.-P. Aubry. A late Paleocene-early Eocene NW European and North Sea magnetobiochronological correlation network; In: *Correlation of the early Paleogene in northwest Europe.* R. W. O'B.Knox , R. M. Corfield and R. E. Dunay, (Eds.). pp. 309-352. Geological Society Special Publication 101 (1996).
4. Berggren, W. A. and D. R. Prothero. Eocene-Oligocene climatic and biotic evolution: an overview. In: *Eocene-Oligocene climatic and biotic evolution.* D. M. Prothero and Berggren, W. A. (Eds.). pp. 1-28. Princeton University Press (1992).
5. Berggren, W. A., D. V. Kent, J. J. Flynn, and J. A. Van Couvering. Cenozoic geochronology, *Geol. Soc. Amer. Bull.* 96:1407-1418 (1985).
6. Bruijn, H. de., R. Daams, G. Daxner-Höck, V. Fahlbusch, L. Ginsburg, P. Mein, J. Morales, E. Heinzmann, D. F. Mayhew, A. J. van der Meulen, N. Schmidt-Kittler and M. Telles-Antunes. Report of the RCMNS working group on fossil mammals, Reisenburg 1990, *Newsl. Strat.* 26: 65-118 (1992).
7. Clemens, W. A., J. A. Lillegraven, E. H. Lindsay and G. G. Simpson. Where, when, and what--a survey of known Mesozoic mammal distribution; In: *Mesozoic mammals.* J. A. Lillegraven, Z. Kielan-Jaworowska and W. A. Clemens. (Eds.). pp. 7-58. University of California Press (1979).
8. Cowie, J. W., W. Ziegler, A. J. Boucot, M. G. Bassett, and J. Remane. Guidelines and statutes of the International Commission on Stratigraphy (ICS). *Cour. Forschungsinst. Senck.* 83: 1-14 (1986).
9. Dashzeveg, D. La faune de mammifères du Paléogène inférieur de Naran-Bulak (Asie centrale) et ses corrélations avec l'Europe et l'Amérique du Nord: *Bull. Soc. Géol. France* 24: 275-281 (1982).
10. Dashzeveg, D. Holarctic correlation of non-marine Paleocene-Eocene boundary strata using mammals. *J. Geol. Soc. London* 145: 473-478 (1988).
11. Dashzeveg, D. Asynchronism of the main mammalian faunal events near the Eocene-Oligocene boundary. *Tert. Res.* 14: 141-149 (1993).
12. Ducrocq, S. Mammals and stratigraphy in Asia: is the Eocene-Oligocene boundary at the right place?". *Com. Rend. Acad. Sci. Paris II* 316: 419-426 (1993).
13. Emry, R. J., S. G. Lucas, L. Tyutkova and B. Wang. The Ergilian-Shandgolian (Eocene-Oligocene) transition in the Zaysan basin, Kazakhstan. *Bull. Carnegie Mus. Nat. Hist.* in press (1997).
14. Fahlbusch, V. Report on the International Symposium on mammalian stratigraphy of the European Tertiary. *Newsl. Strat.* 5: 160-167 (1976).
15. Gunnell, G. F., W. S. Bartels, and P. D. Gingerich. Paleocene-Eocene boundary in continental North America: Biostratigraphy and geochronology, northern Bighorn basin, Wyoming, *New Mex. Mus. Nat. Hist. Sci. Bull.* 2: 137-144 (1993).
16. Hooker, J. J. The sequence of mammals in the Thanetian and Ypresian of the London and Belgian basins. Location of the Paleocene/Eocene boundary, *Newsl. Stratig.* 25:

75-90 (1991).

17. Hooker, J. J. British mammalian paleocommunities acrossd the Eocene-Oligocene transition and their environmental implications. In: *Eocene-Oligocene climatic and biotic evolution*. D. M. Prothero, and W. A. Berggren (Eds.). pp. 495-515. Princeton University Press (1992).

18. Hooker, J. J. 1996, Mammalian biostratigraphy across the Paleocene-Eocene boundary in the Paris, London and Belgian basins .In: *Correlation of the early Paleogene in northwest Europe*. R. W. O'B.Knox , R. M. Corfield and R. E. Dunay, (Eds.). pp. 205-218. Geological Society Special Publication 101 (1996).

19. Lofgren, D. L. The Bug Creek problem and the Cretaceous-Tertiary transition at McGuire Creek, Montana, *Univ. Calif. Pub. Geol. Sci.* 140: 1-185 (1995).

20. Lopatin, A. V. Stratigrafiya i melkiye mlyekopitayushchiye Aralskoi svity Altynshokysu (Severnoye Priaralye), *Strat. Geol. Korrel.* 4: 65-79 (1996).

21. Lucas, S. G. Vertebrate biochronology of the Cretaceous-Tertiary boundary, San Juan Basin, New Mexico. *28th Int. Geol. Congr. Field Trip Guideb.* T120: 47-51 (1989).

22. Lucas, S. G. Fossil mammals and the Paleocene/Eocene Series boundary in Europe, North America and Asia. In: *Paleocene/Eocene climatic and biotic events*. M. P. Aubry, S. G. Lucas and W. A. Berggren (Eds.). in press. Columbia University Press (1997).

23. Lucas, S. G., E. G. Kordikova and R. J. Emry. Oligocene stratigraphy, sequence stratigraphy and mammalian biochronology north of the Aral Sea, western Kazakhstan, *Bull. Carnegie Mus. Nat. Hist.*, in press (1997).

24. Mein, P. Résultats du groupé de travail des vertébrés. *Rep. Act. RCMNS Work. Groups (1971-1975)*:: 78-81, Bratislava (1975).

25. Premoli-Silva, I. and D. G. Jenkins. Decision on the Eocene-Oligocene boundary stratotype, *Episodes* 16: 379-382 (1993).

26. Premoli-Silva, I., R. Coccioni, and A. Montanari, eds. *The Eocene-Oligocene boundary in the Marche-Umbria basin (Italy)*. Ancona, IUGS (1988).

27. Prothero, D. R. Chadronian (early Oligocene) magnetostratigraphy of eastern Wyoming: implications for the Eocene-Oligocene boundary. *J. Geol.* 93: 555-565 (1985).

28. Prothero, D. R. and C. C. Swisher, III. Magnetostratigraphy and geochronology of the terrestrial Eocene-Oligocene transition in North America. In: *Eocene-Oligocene climatic and biotic evolution*. D. M. Prothero, and W. A. Berggren (Eds.). pp. 46-73. Princeton University Press (1992).

29. Russell, D.E. and R. Zhai. The Paleogene of Asia: Mammals and stratigraphy: *Mém. Mus. Nat. Hist. Nat. C Sci. Terre* 52:1-488 (1987).

30. Russell, L. Cretaceous non-marine faunas of northwestern North America. *Roy. Ont. Mus. Life Sci. Contrib.* 61: 1-24 (1964).

31. Russell, L. Mammalian faunal succession in the Cretaceous System of western North America. *Geol. Assoc. Canada Spec. Pap.* 13: 137-161 (1975).

32. Shevyryeva, N. S. O vozraste gryzunov (Rodentia, Mammalia) Buranskoi svity Zaisanskoi vpadiny (vostochniyi Kazakhstan), *Strat. Geol. Korryel.* 3: 73-82 (1995).

33. Stehlin, H. G. Remarques sur les faunules de mammifères des couches éocènes et oligocènes du Bassin de Paris. *Bull. Soc. Géol. Fr.* 9: 488-520 (1909).

34. Steininger, F. F., R. L. Bernor, and V. Fahlbusch. European Neogene marine/continental chronologic correlations. In: *European Neogene mammal chronology*. E. H. Lindsay, V. Fahlbusch, and P. Mein (Eds.). pp. 15-46. Plenum

Press,New York (1990).

35. Tedford, R. H., M. F. Skinner, R. W. Fields, J. M. Rensberger, D. P. Whistler, T. Galusha, B. E. Taylor, J. R. Macdonald and S. D. Webb. Faunal succession and biochronology of the Arikareean through Hemphillian interval (late Oligocene through earliest Pliocene Epochs) in North America. In: *Cenozoic mammals of North America*. M. O. Woodburned (Ed.).pp. 153-210. University of California Press, Berkeley (1987).

36. Tedford, R. H., J. B. Swinehart, C. C. Swisher III, D. R. Prothero, S. A. King and T. E. Tierney. The Whitneyan-Arikareean transition in the High Plains. In: *The terrestrial Eocene-Oligocene transition*. D. R. Prothero and R. J. Emry (Eds.). pp. 312-334. Cambridge University Press (1996).

37. Ting, S. Paleocene and early Eocene land mammal ages of Asia, *Bull. Carn. Mus. Nat. Hist.*, in press (1996).

38. Tong, Y., S. Zheng and Z. Qiu.Cenozoic mammal ages of China. *Vert. PalAs.* 33: 290-314 (1995).

39. Van Valen, L. The beginning of the Age of Mammals. *Evol. Theory* 4: 45-80 (1978).

40. Vislobokova, I. A. O vozraste fauny Shand-Gol Mongolii i evolyutsii fauny mlyekopitayushchikh tsentralnoi Azii v Oligotsene. *Strat. Geol. Korrel.* 4: 55-64 (1996).

41. Wang, B. The Chinese Oligocene: a preliminary review of mammalian localities and local faunas; In: *Eocene-Oligocene climatic and biotic evolution*. D. R. Prothero and W. A. Berggren (Eds.).pp. 529-547. Princeton University Press (1992).

42. Williamson, T. E. The beginning of the age of mammals in the San Juan Basin, New Mexico: Biostratigraphy and evolution of Paleocene mammals of the Nacimiento Formation: *New Mex. Mus. Nat. Hist. Sci. Bull.* 8: 1-141 (1996).

43. Wood, H.E., II, R.W. Chaney, J. Clark, E. H. Colbert, G. L. Jepsen, J. B.Reeside, Jr., and C. Stock. Nomenclature and correlation of the North American continental Tertiary. *Geol. Soc. Amer. Bull.* 52: 1-48 (1941).

44. Woodburne, M. O. and C. C. Swisher, III. Land mammal high-resolution geochronology, intercontinental overland dispersals, sea level, climate, and vicariance, *SEPM Spec.Pub.* 54: 335-364 (1995).

45. Yanovskaya, N. M., Ye. N. Kurochkin, and Ye. V. Devyatkin. Mestonakhozhdeniye Ergilin-Dzo--stratotip nizhnevo Oligotsena v yugo-vostochnoi Mongolii, *Trudy Sovm-Mongol. Paleont. Eksped.* 4: 14-33 (1977).

Proc. 30ᵗʰ Int'l. Geol. Congr., Vol. 11 , pp. 201-211
Wang Naiwen and J. Remane (Eds)
© VSP 1997

A candidate section for the Lower - Middle Pleistocene Boundary (Apennine Foredeep, South Italy)

NERI CIARANFI *, ASSUNTA D'ALESSANDRO * & MARIA MARINO *

* Department of Geology and Geophysics, University of Bari, Italy

Abstract

In the southernmost part of the Bradano Trough, in Basilicata (South Italy), there is a well exposed, continuous succession of muds and muddy silts, about 400 metres thick: on the basis of nannofossil biostratigraphic analyses and according to preliminary magnetostratigraphic results, it has been referred to the upper part of the Lower Pleistocene and to the lower part of the Middle Pleistocene. The composite section of Montalbano Jonico has been referred to the top of the "large" *Gephyrocapsa* Zone, to the "small" *Gephyrocapsa* Zone and to the lower part of the *Pseudoemiliania lacunosa* Zone. The boundaries between the above-mentioned biozones have been recognized close to geomagnetic reversals corresponding to the top of the Jaramillo subchron and to the Matuyama-Brunhes boundary respectively. Faunal composition together with taphonomic and sedimentological observations, have given basic information about the bathymetric evolution of the Montalbano Jonico basin pointing out a general regressive trend that includes at least five cycles of fifth order sea level rise and fall, in the interval referred to the *P. lacunosa* Zone. The composite section of Montalbano Jonico seems to contain two possible intervals useful for the GSSP of the Lower-Middle Pleistocene Boundary: a) the lowermost interval includes the top of the Jaramillo subchron and the small *Gephyrocapsa/P. lacunosa* zonal boundary; b) the topmost one includes the Matuyama/Brunhes boundary.

Keywords: Stratigraphy, Marine Pleistocene, South Italy

INTRODUCTION

The GSSP for the Lower-Middle Pleistocene Boundary has never been formally established following the statements of the International Stratigraphic Guide [28]. As reported by Cita & Castradori [10; 11], the boundary between these two Pleistocene subdivisions should be placed close to the top of the Jaramillo subchron, which is near the isotope stage 25 and/or the *small Gephyrocapsa/Pseudoemiliania lacunosa* nannofossil zonal boundary. Alternatively, other researchers [23, 25] suggest to place the same boundary close to the Brunhes-Matuyama reversal.

In the Montalbano Jonico area, southwestern part of the Bradano Trough, one of the most recent on-land marine Quaternary successions is well preserved. Its record is continuous and it seems to contain both the base and the top of the Jaramillo, as well as the Matuyama-Brunhes Boundary (Channell, pers. comm., 1994), according to the stratigraphical position of nannofossil zonal boundaries [7, 8, 9]. The Montalbano Jonico Section shows the most relevant characteristics required [12, 28] to be candidated for the selection of the GSSP of the Lower-Middle Pleistocene Boundary. The submission of the section has been also suggested by the Quaternary Working Group of the Italian Commission on Stratigraphy [10; 11] and by Van Couvering [31].

GEOLOGICAL SETTING

All the stratigraphic sections used to establish the marine Quaternary Stratotypes are located on-land areas along the eastern margin of the Apennine chain (Fig. 1a): the reason can be attributed to the peculiar geodynamic and paleogeographic evolution of foredeep basins. In South Italy, such a basin extends from Molise to Basilicata regions and expands into the present-day Taranto Gulf; it is located between the external border of the Apennine fold-and-thrust belt and the internal margin of the Apulian Foreland [3, 5; 6; 22].

This basin rapidly evolved into a filled-up trough, called Bradano Trough (Fig. 1b), that represents one of the main physiographic elements of South Italy. The trough, which was formed in the Early Pliocene, deeply evolved during the Pliocene and the Early Pleistocene. Its inner border, deformed by polyphasic active thrusts, moved towards NE involving Cenozoic Apennine Units and Pliocene Foredeep ones. The outer margin spread gradually over the Meso-Cenozoic Apulian Foreland Units [3, 5].

During the Quaternary, the Bradano Trough was characterized by the continuous deposition of a thick succession along the depocentral areas. Sedimentation went on for a longer time in southern Basilicata, thus the Montalbano Jonico record represents the most recent on-land succession discovered in Italy so far.

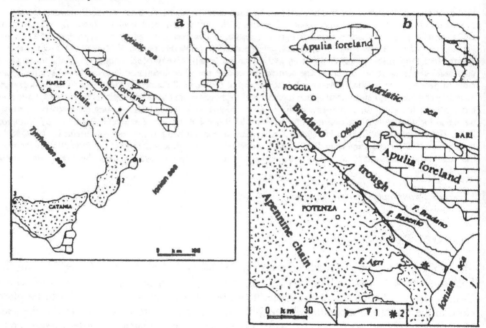

Figure 1. a - Location of stratotype sections in South Italy. 1 Vrica: 2 S.Maria di Catanzaro; 3 Ficarazzi; 4 Montalbano Jonico. **b** - 1 Buried allochtonous outer front; 2 Montalbano Jonico section.

STRATIGRAPHY

The Montalbano Jonico section consists of coarsening upward muddy deposits belonging to the "Argille subappennine" Formation. A composite section, over 400 m thick, has been reconstructed in the field by means of selected stratigraphic sections including some volcaniclastic layers (Figs. 2, 3). The correlation has been made possible through the presence of peculiar fossil assemblages that characterize each volcaniclastic layer, therefore these layers can be used as guide horizons. Moreover, magnetostratigraphic studies and biostratigraphic analyses on nannoflora assemblages have improved the reconstruction of this succession [7; 8; 9; 13; 20]. The volcaniclastic layers (V1-V9) - made of pure ashes, sands rich in volcanic minerals and pumice clasts - have been referred to an alkaline undersatured volcanism, probably coming from a South Italy volcanic source [8].

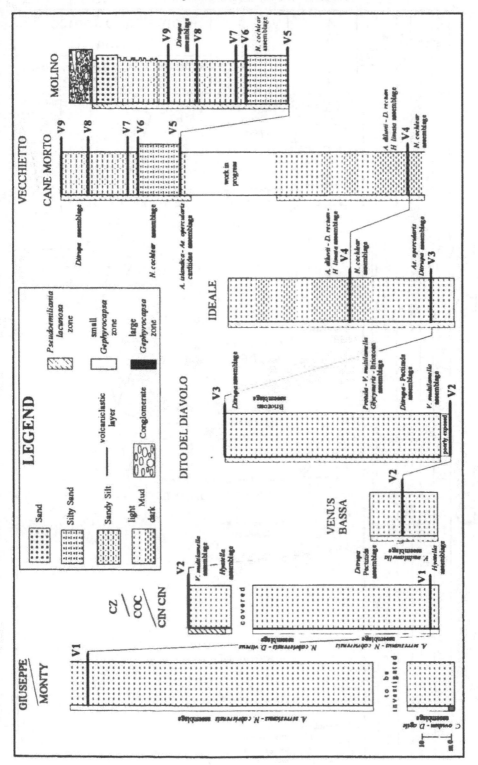

Figure 2. Selected stratigraphic sections in the Montalbano Jonico area: the correlations have been obtained using volcaniclastic layers and fossil assemblages.

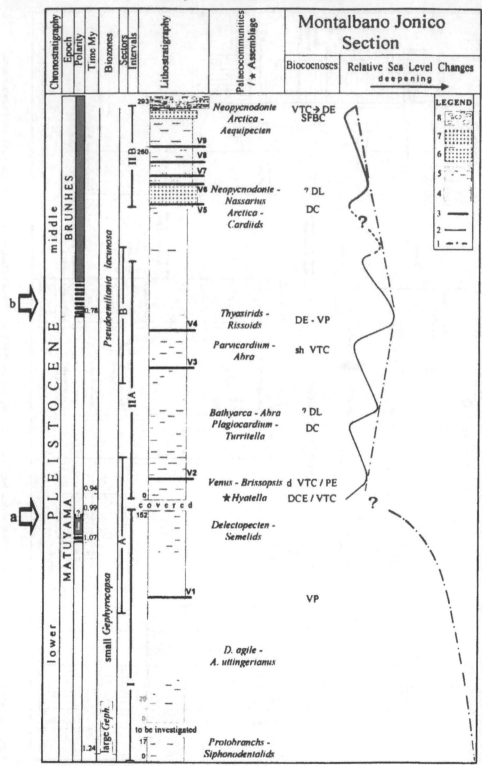

The lower 150 metres of the succession are characterized by hemipelagic muds and include the lowest volcaniclastic layer V1. The fossil remains that occur in the field mainly belong to *Dentalium agile, Aphorrais uttingerianus, A. serresianus* and *Nassarius cabrierensis*.

Muddy sediments, including the volcaniclastic layers V2 - V4, follow upward for a thickness of about 200 m (Fig. 3). These deposits contain dispersed macrofossils as well as skeletal concentrations, which are mainly referable to primary biogenic concentrations dominated by bryozoans (below V3) or bivalves (near V2 and V4).

The upper part of the succession, about 70 m thick, consists of silts and sands, in which the V5 - V9 volcaniclastic layers are included. Winnowing and current concentrations, replaced upsection by proximal tempestites, occur in addition to biogenic concentrations. These sediments represent the topmost regressive part of a marine cycle.

In paraconformity on this record, shelf sands and conglomerates crop out for a thickness of about 10 m [18]. A continental sandy-conglomeratic body covers the marine deposits.

NANNOFOSSIL BIOSTRATIGRAPHY

Two selected sectors of the composite Section (Fig. 3) have been investigated in detail: *Sector A* including the inferred Jaramillo Subchron, and *Sector B* in which the Matuyama-Brunhes Boundary has been preliminarily recorded.

Samples for calcareous nannofossil biostratigraphy have been analyzed every one metre. Quantitative studies have been restricted to the sectors *A* and *B* (Fig.4), whereas the other intervals have been qualitatively examined. Analyses have been performed on smear slides with a light microscope at 1000X magnification. An average of 400 fields of view (= 8 mm^2) has been investigated in order to determine the significant assemblage of each sample and to recognize the abundance patterns of the marker taxa (gephyrocapsids, helicospherids, *Reticulofenestra asanoi* Sato and Takayama). Taxonomic concepts about *Gephyrocapsa* spp. follow the classification proposed by Gartner [16], Rio [26], Rio *et al.* [27] and Raffi *et al.* [24]. The abundances of normal sized *Gephyrocapsa*, large *Gephyrocapsa* and *R. asanoi* have been plotted as number of specimens/mm^2; whereas, the abundance of *Gephyrocapsa* sp. 3 and *Helicosphaera sellii* has been plotted as percentage related to the total gephyrocapsids (>4 μm) and helicoliths respectively, counted on 8 mm^2. The abundance of small *Gephyrocapsa* has been represented as mean number of specimens per field of view, based on a count of this morphotype on 20 fields of view.

Three biozones have been recognized according to the scheme of Rio *et al.* [27]: large *Gephyrocapsa* Zone p.p., small *Gephyrocapsa* Zone and *Pseudoemiliania lacunosa* Zone p.p.

Sector A

The small *Gephyrocapsa/P. lacunosa* zonal boundary has been defined on the basis of the first occurrence of *Gephyrocapsa* sp. 3 in the upper part of this sector; the end acme of small *Gephyrocapsa* and the re-entry of normal sized *Gephyrocapsa* occur very close to this zonal boundary. The presence of a covered stratigraphical interval (see Fig. 4) just below these bioevents may suggest a slightly lower position of the small *Gephyrocapsa/P. lacunosa* zonal boundary.

Low abundances of normal sized *Gephyrocapsa* and high abundances of small *Gephyrocapsa*

Figure 3. Composite section of Montalbano Jonico. **a** - Lowermost location of the GSSP possible for the Lower-Middle Pleistocene Boundary. **b** - Topmost location of the GSSP possible for the Lower-Middle Pleistocene Boundary. The age of geomagnetic reversals is from Shackleton *et al.* [29]; the reported ages of nannofossil zonal boundaries are from Raffi *et al.* [24] and Castradori [4]. Changes in water depth have been inferred by the biocoenotic assignment of the palaeocommunities. Biocoenoses and their relative letter code: VP=Bathyal Mud, DC=Coastal Detritic Bottoms. VTC-Terrigenous Mud. PE. Heterogeneous Community. DI. Shelf-Edge Detritic Bottoms. DE=Muddy Detritic Bottoms, DCE=DC-DE ecotone; sh shallow facies. d−deep facies. 1) fourth order cycle; 2) fifth order cycle; 3) volcaniclastic layers; 4) mud; 5) sandy silt; 6) silty sand; 7) sand; 8) conglomerate.

characterize the assemblages of the small *Gephyrocapsa* Zone. Moreover, the "base acme" and the "top acme" of *R. asanoi* occur in the dominance range of small *Gephyrocapsa* according to several authors [4, 14; 19]. The distribution of *R. asanoi* results to be slightly variable probably due to different taxonomic concepts about the species [discussion in 20]. The base of *P. lacunosa* Zone is characterized by the re-entry of normal sized *Gephyrocapsa* and by a lower abundance of small *Gephyrocapsa.* The re-entry of normal sized *Gephyrocapsa* is largely represented by *Gephyrocapsa*

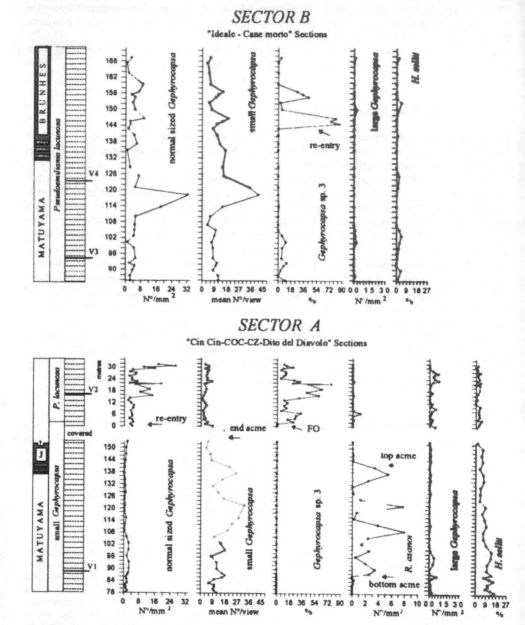

Figure 4. Abundance patterns of main nannofossil markers in the *Sectors A* and *B* of the Montalbano section, see text for further explanations.

sp. 3, the percentage of which varies from 6% to 83%. The re-entry of medium sized *Gephyrocapsa* seems to be associated with isotope stage 25 [4], with stage 27 [24, 30] or with stage 28-29 [24] reflecting a migratory event from lower to mid and high-latitudes [24].
Several specimens of large *Gephyrocapsa* and *H. sellii* are reworked in the *Sector A*.

Sector B

The qualitative analyses carried out on several samples collected between the *Sectors A* and *B* and above the *Sector B*, put in evidence the discountinuous presence of *Gephyrocapsa* sp. 3; this is in agreement with the distribution known for this marker in the Mediterranean area [4], where the same species desappears at 584 ky [4]. On the basis of quantitative analyses, *Gephyrocapsa* sp. 3 results to be rare or absent in the basal portion of the *Sector B* and becomes abundant in the upper part of the section (Fig. 4). The discontinuous distribution of *Gephyrocapsa* sp. 3 in the Montalbano succession, possibly amplified by high accumulation rates and variable inorganic input, prevents an unambiguous interpretation of the abundance pattern of the species as well as a straight comparison with the few quantitative data available in literature. The biostratigraphic meaning of the "re-entry" of *Gephyrocapsa* sp. 3, observed in the *Sector B* (Fig. 4), needs still to be valued with further analyses.

PALAEOCOMMUNITIES AND SEA LEVEL CHANGES

Since from the preliminary surveys, marked sea-level changes have been inferred by the vertical distribution of the fossil assemblages in the Montalbano section. Almost all macrofossils belong to still-extant taxa related to communities inhabiting different biotopes from the upper slope to the inner shelf. Freshly exposed surfaces of selected sections have been examined in detail to identify the original benthic palaeocommunities and to assign them to the biocoenoses [*sensu* 21] of the past - here named "associations" - which are believed to play the same ecological role as the modern ones [2]. The bionomic approach made possible to obtain reliable information on the changes through time of the main environmental parametres, primarily bathymetry.
The studied record can be divided into two intervals (Fig. 3): the lower one (I) is characterized by epibathyal fossil communities, the upper one (IIA, IIB) is characterized by sublittoral palaeocommunities whose stratigraphical distribution reveals five (possibly six) sea-level fluctuations.

Interval I.

This interval consists of poorly fossiliferous trace-laminated and massive muds. Trace fossils are mostly represented by *Chondrites*. The three identified palaeocommunities are equivalent to facies of the modern VP biocoenosis and reveal a shallowing trend, from a depth around 500 m to the transitional belt up to the shelf (Fig. 3). Furthermore, the compositional and structural changes of the palaeocommunities, trace fossil assemblages together with the observed taphonomic signatures reflect variations in oxygen content and in edaphic parametres (basically soupiness of the sea-floor); besides, they reveal the occurrence of numerous blanketing events by resuspended muds.

Interval II.

This interval has been almost completely measured and sampled metre-by metre or, whenever possible, in more detail. All the palaeocommunities are comparable to modern biocoenoses inhabiting the sublittoral zone. In most cases the palaeocommunities seem to replace each other because of the gradual variation of some environmental parametres, i.e. bathymetry, bottom cohesivity, sedimentation rate, oxygen concentration.
Some depth-related palaeocommunities have been selected in order to define the most relevant deepening-shallowing cycles [14] (Fig. 3). Each cycle includes a number of subcycles showing smaller-magnitude depth changes, delineated by subtle faunistic and taphonomic variations. The cyclicity is partly masked by colonization events that possibly were triggered by short-term

processes, such as weak rises of water energy (able to remove only a fluid interface or just the finest material), episodic increase of turbidity and mud mass deposition.

The sub-interval IIA is a poorly fossiliferous, muddy sediment characterized by the intercalations of loosely to densely packed fossil concentrations forming pavements (mostly made of pectinid shells and/or tubeworms), beds (mostly bryozoan colonies), clumps (*Neopycnodonte* or *Ditrupa*) and thin lenses. From a genetic point of view, beds and clumps represent intrinsic biogenic concentrations [17], whereas the pavements are mainly sedimentologic concentrations due to the removal of finest sediment.

At the base of this sub-interval (Fig. 3), sedimentological concentrations (*Hyatella* is a characteristic component, with *Glossus, Ditrupa, Pseudamussium*) and few erosional surfaces indicate episodic increasing in the bottom water energy. Taphonomic features and ecological meaning suggest weak displacements and mixing of skeletons that were rapidly buried by muddy tempestites. The seafloor is inferred to be located close to the maximum storm wave base. Upwards, two palaeocommunities (*Plagiocardium-Turritella* and *Parvicardium-Abra*) - equivalent to the DC biocoenosis and a shallow facies of the VTC biocoenosis - suggest similar relatively low depth bottoms. The maximum depth of each cycle is recorded by the *Brissopsis-Venus, Bathyarca-Abra, Thyasirids-Rissoids* palaeocommunities.

The sub-interval IIB (Fig. 3) consists of sandy-silts intercalated with silty muds and fine sands. The uniformly distributed faunal remains are generally more abundant than in IIA, whereas the primary biogenic concentrations are rare. The mixed biologic-sedimentologic concentrations, mostly generated by short-term processes affecting a bottom below the storm wave base, are better represented. The *Arctica-Cardiids* (DC association) and the *Neopycnodonte-Nassarius* palaeocommunities (DE association) indicate a deepening trend of the seafloor from 30-40 m to around 100 m. Upwards, the minimum depth is marked by a SFBC association (*Arctica-Aequipecten* palaeocommunity) that testifies a shallowing of the bottom, up to a depth of about 10 m. The last recorded transgression is testified by the presence of numerous *Brissopsis* and by *Neopycnodonte* clumps.

The envelope of the maximum depth values of the fifth order cycles reveal a larger-scale cycle.

CONCLUSIONS

In the southernmost on-land area of the Bradano Trough there is a well exposed continuous succession of muds and muddy silts, about 400 metres thick, which, on the basis of nannofossil biostratigraphy and preliminary magnetostratigraphic studies, can be referred to the upper part of the Lower Pleistocene and to the lower part of the Middle Pleistocene.

In the Montalbano Jonico composite section two potential locations for the GSSP of the Lower-Middle Pleistocene Boundary are available according to the basic features required in "Guidelines for Boundary Stratotypes" [12]; in fact the section is represented by sedimentary continuous hemipelagites deposited in a well known paleogeographic and tectonic environment. Furthermore, the outcrop is easily accessable and shows several marker horizons (volcaniclastic layers); it contains well preserved and diversified faunal and floral assemblages. Lithologies are prone to isotopic, geochronometric and detailed magnetostratigraphic analyses.

The first potential boundary is located in the *Sector A* (Fig.3) close to the top of the Jaramillo Subchron (990 ky) [29] near the zonal boundary between the small *Gephyrocapsa* and the *Pseudoemiliania lacunosa* Zones (940 ky) [4], and it can be correlated with the isotope stage 25 [4]. A covered interval, just above the top of the Jaramillo, should lower the top of the small *Gephyrocapsa* Zone. Further field investigations are still in progress to better describe this interval in order to locate precisely the biostratigraphic zonal boundary within the Montalbano Section. The second position suggested for the Lower-Middle Pleistocene GSSP is located in the *Sector B* (Fig. 5), close to the Matuyama-Brunhes Boundary (720 ky) [29].

The whole Montalbano section was deposited between 1.24 My [24] - corresponding to the LAD of large *Gephyrocapsa* - and an age not younger than about 600 ky, for the presence of *Gephyrocapsa* sp. 3 at the top of the section. It is likely to represent the regressive part of a third-order cycle. A relevant tectonic pulse seems to be the main cause of the general shallowing. Positive pulses are strictly linked to changes in the geodinamic regime already suggested for the beginning of the filling-up phase in the Bradano Trough evolution [5].

The suggested fourth-order cycles are believed to be originated by tectonoeustatic causes. The boundary between the two recognized cycles is probably connected with the last warm event prior to the "glacial" Pleistocene, i.e. isotope stage 25.

The five, possibly six, fifth order cycles between 940 ky and about 600 ky seem to be related to an eustatically controlled cyclicity, with a good accordance to the already known Middle Pleistocene cyclicity. The similar time-span of each cycle suggests possible astronomical relationships.

The great thickness of the Montalbano section reflects a high average rate of accumulation mainly depending on events of high sedimentation rate due to denudation of the land areas surrounding this basin, to a high-frequence storm transport of the resuspended mud, and to a very weak erosion of the seafloor.

Figure 5. Badlands cropping out west of Montalbano Jonico, including the "Ideale section" (see Fig. 2). V3 and V4 are two prominent volcaniclastic marker-beds; b indicates the location of the Matuyama-Brunhes reversal.

Acknowledgements

We are indebted to M.B. Cita for advice and encouragment, to D. Castradori and J. Remane for the accuracy of reviewing and for suggestions.
We are very grateful to our young colleagues P. Maiorano, R. Sbarra and D. Soldani for help in the field and in the laboratory.
We also thank Regione Basilicata and Comune di Montalbano Jonico for financial support and facilities offered respectively.

REFERENCES

1. P. Ambrosetti, C. Bosi, F. Carraro , N. Ciaranfi, M. Panizza, G. Papani, L. Vezzani and A. Zanferrari. *Neotectonic map of Italy*, Quad. Ric. Scient., 4, Roma (1987).

2. M. P. Bernasconi and E. Robba - *Molluscan palaeoecology and sedimentological features: an integrated approach from the Miocene Meduna section, Northern Italy*. Palaeogeogr., Palaeoclim., Palaeoecol., 100, 267-290 (1993).

3. R. Casnedi, U. Crescenti and M. Tonna. *Evoluzione dell'avanfossa adriatica meridionale nel Plio-Pleistocene, sulla base di dati di sottosuolo*. Mem. Soc. Geol. It., 24, 243-260 (1982).

4. D. Castradori. *Calcareous nannofossil biostratigraphy and biochronology in eastern Mediterranean deep-sea cores*. Riv. It. Paleont. Strat., 99, 107-126 (1993).

5. N. Ciaranfi, M. Maggiore, P. Pieri, L. Rapisardi, G. Ricchetti and N. Walsh. *Considerazioni sulla neotettonica della Fossa bradanica*. P.F.Geodinamica, 251, 73-95, Napoli (1979).

6. N. Ciaranfi , M. Guida, G. Iaccarino, T.Pescatore, P. Pieri, L. Rapisardi, G. Ricchetti, I. Sgrosso, M. Torre, L. Tortorici and E. Turco, R. Scarpa, M. Cuscito, I. Guerra, G. Iannaccone, G.F. Panza and P. Scandone. *Elementi sismotettonici dell'Appennino meridionale*. Boll. Soc. Geol.It., 102, 201-222 (1983)

7. N. Ciaranfi, A. D'Alessandro, M. Marino and L. Sabato. *La successione argillosa infra e mediopleistocenica della parte sudoccidentale della Fossa bradanica: la Sezione di Montalbano Jonico in Basilicata*. In Guida alle Escursioni del 77° Cong.Naz. della S.G.I.; Quad.Bibl.Prov. Matera, 15, 117-156 (1994).

8. N. Ciaranfi, M. Marino, L. Sabato, A. D'Alessandro and R. De Rosa. *Studio geologico-stratigrafico di una successione infra e mesopleistocenica nella parte sud-occidentale della Fossa bradanica (Montalbano Jonico, Basilicata)*. Boll. Soc. Geol. It., 115, 379-391 (1996).

9. N. Ciaranfi. and M. Marino. *A potential type-locality for the Lower-Middle Pleistocene Boundary Stratotype (Apennine Foredeep, Southern Italy)*. Terra Nova, INQUA XIV Int. Congr., Abstract, Berlin (1995).

10. M.B. Cita & D. Castradori. *Rapporto sul workshop "Marne sections from the Gulf of Taranto (Southern Italy) usable as a potential Stratotypes for the GSSP of the Lower, Middle and Upper Pleistocene" (29 settembre - 4 ottobre 1994)*. Boll.Soc.Geol.It., 114, 319-336 (1995a).

11. M.B. Cita & D. Castradori. *Workshop on marine sections of Gulf od Taranto (Southern Italy) usable as a potential Stratotypes for GSSP of the Lower, Middle and Upper Pleistocene* (Bari, Italy, Sept 29-Oct.4). Il Quaternario, 7, 677-692 (1995b).

12. J. W. Cowie. *Guidelines for Boundary Stratotypes*. Episodes, 9, 78-82 (1986).

13. A. D'Alessandro, L. Sabato, M. Marino and N. Ciaranfi. *Environmental evolution of an Early-Middle Pleistocene succession (Montalbano Jonico, Southern Italy)*. Terra Nova, INQUA XIV Int.Congr., Abstract, 57, Berlin, August 1995.

14. A. D'Alessandro, A. Di Bernardi and N. Ciaranfi. *Environmental evolution of an Early-Middle Pleistocene succession of internal foredeep near Montalbano Jonico (MT)*. Congr. Soc. Paleont. It., Abstract, Settembre 1996, Parma.

15. J.V. Firth and M. Isiminger-Kelso. *Pleistocene and Oligo-Miocene calcareous nannofossils from the Sumisu Rift and Izu-Bonin Forearc Basin*. In B. Taylor, K. Fuijoka et al., 1992. Proc. ODP, Scient. Results, 126, 237-262 (1992).

16. S. Gartner. *Correlation of Neogene planktonic Foraminifera and Calcareous Nannofossil zones*. Trans. Gulf Coast Assoc. Geol. Soc., 19, 585-599 (1977).

17. S.M. Kidwell, F.T. Fursich and T. Aigner. *Conceptual framrwork for the analysis and classification of fossil concentrations*. Palaios, 1, 228-338 (1986).

18. F. Loiacono, M. Moresi, A. D'Alessandro and N. Ciaranfi. *Caratteri di facies della Sezione stratigrafica infra-mesopleistocenica di Montalbano Jonico (Fossa bradanica)*. Riunione Gruppo Sedimentologia del CNR Atti (A. Colella ed.), 172-174. Catania, 10 ottobre 1996.

19. P. Maiorano, M. Marino and S. Monechi. *Pleistocene calcareous nannofossil high resolution biostratigraphy of Site 577, Northwest Pacific Ocean*. Palaeopelagos, 4, 119-128 (1994), Roma.

20. M. Marino. *Quantitative calcareous nannofossil biostratigraphy of the Lower-Middle Pleistocene Montalbano Jonico Section (Southern Italy)*. Palaeopelagos Journal, 6, in press.

21. J.-M. Pérès and J. Picard. *Nouveau manuel de bionomie benthique de la Mer Méditerranée*. Recueil Trav. St. Mar. Endoume, 31, 137 pp. (1964).

22. P. Pieri, L. Sabato and M. Tropeano. *Significato geodinamico dei caratteri deposizionali e strutturali della Fossa bradanica nel Pleistocene*. Mem. Soc. Geol. It. 51, 501-515 (1996).

23. B. Pillans, S.T. Abbot, A.G. Beu and R.M. Carter. *A possible Lower/Middle Pleistocene boundary stratotype in Wanganui Basin, New Zealand*, Abstract. Int. Union Quat. Res. (Beijing), 281, 1991.

24. I. Raffi, J. Backman, D. Rio and J. Shackleton. *Plio-Pleistocene nannofossil biostratigraphy and calibration to oxygen isotope stratigraphies from DSDP Site 607 and ODP Site 677*. Paleoceanography, 8, 387-408 (1993).

25. G. M. Richmond. *The INQUA - approved provisional Lower-Middle Pleistocene boundary.* In The early Middle Pleistocene in Europe. Turner (Ed.), 319-327. Balkema, Rotterdam (1996).
26. D. Rio. *The fossil distribution of Coccolithophore Genus Gephyrocapsa Kamptner and related Plio-Pleistocene chronostratigraphic problems.* In W.L. Prell, J.V. Gardner *et al.*, Init. Repts. DSDP, 68, 325-343 (1982).
27. D. Rio, I. Raffi and G. Villa. *Pliocene-Pleistocene calcareous nannofossil distribution patterns in the western Mediterranean.* In K. Kastens, J. Mascle *et al.*, Proc. ODP, Scient. Results, 107, 513-533 (1990).
28. A. Salvador (Ed.). *International Stratigraphic Guide. A guide to stratigraphic classification, terminology and procedure.* Second Edition, Boulder, CO, The Geol. Soc. of America (1994).
29. J. Shackleton, A. Berger and W.R. Peltier. *An alternative astronomical calibration on the lower Pleistocene timescale based on ODP Site 677.* Trans.Royal Soc. Edinburg: Earth Scien., 81, 251-261 (1990).
30. R. Sprovieri. *Pliocene-Early Pleistocene astronomically forced planktonic Foraminifera abundance fluctuations and chronology of Mediterranean calcareous plankton bio-events.* Riv. It. Paleont. Strat., 99, 371-414 (1993).
31. J. Van Couvering. *Setting Pleistocene marine stages.* Geotimes, 40, 10-11 (1995).

Proc. 30ᵗʰ Int'l. Geol. Congr., Vol. 11 , pp. 213-230
Wang Naiwen and J. Remane (Eds)
© VSP 1997

The Plio-Pleistocene Diatom Record from ODP Site 797 of the Japan Sea

ITARU KOIZUMI and AKIHIRO IKEDA

Division of Earth and Planetary Sciences, Graduate School of Sciences,
Hokkaido University, Sapporo, 060 JAPAN

Abstract

The Plio-Pleistocene diatom record of the ODP Site 797, where is located at the critical position of water masses, can be divided into four sections: (1) the latest Pliocene (2.0-1.6 Ma) where is low abundance with minor periodic fluctuations; (2) the early Pleistocene (1.6-1.1 Ma) where is characterized by the high frequency oscillations and the stepwise spike; (3) the middle Pleistocene (1.1-0.5 Ma) in which moderate abundance and highly complicated fluctuations are seen as compared to the earlier and later intervals; and (4) the late Pleistocene (0.5-0.0 Ma) where the variations in the abundance are dominated by the 100,000-year cycle.

The Maximum Entropy Method (MEM) power spectral density (MEM-PSD) analysis were performed for diatom abundance during the last 1,600 kyr. The predominant period within the range of 105.3-112.2 kyr period, which is characterized by the glacial-interglacial cycle, corresponds to the orbital eccentricity since 1,100 kyr ago. The 2nd-order periods ranging from 40.0 kyr to 42.8 kyr cycle correspond to the obliquity (tilt) band variations. The periods of 28.6 kyr in the early Pleistocene and the 31.0 kyr in the middle Pleistocene correspond to the 29 kyr period in nonlinear interactions of the 100 kyr of eccentricity and the 41 kyr periods of tilt. The 18.3 kyr cycle in the middle-late Pleistocene corresponds to the precession band variations.

Keywords: Diatom, Japan Sea, ODP Site 797, spectral analysis, Milankovitch Cycles

INTRODUCTION

The Japan Sea is a semi-enclosed marginal sea located on the eastern end of the Asian continent with its eastern margin bounded by Japanese Island Arc (Fig. 1). The sea is connected with the East China Sea (ECS) to the southwest, Pacific Ocean to the east, and the Okhotsk Sea to the north through shallow straits. It has been responding as a resonance box to the high frequency changes of global surface conditions in a very sensitive through the narrow and shallow sill depth of the Tsushima Strait (130 m deep), Tsugaru Strait (130 m deep) and Soya Strait (55 m depth). At present, the Tsushima Warm Current (TWC), a branch of the Kuroshio Current, is the only current flowing through the Tsushima Strait into the Japan Sea. The current flows along the western margin of

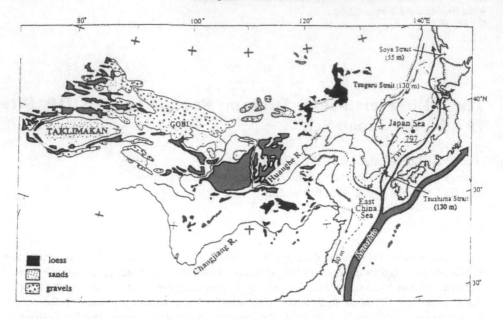

Figure 1. A Locality map showing location of studied core in the Japan
Sea and distribution of loess and major deserts in the central to east Asia.
The modern circulation in the Japan Sea, East China Sea, and Yellow Sea are
shown as arrows. Bathymetric contour of -80m in the East China Sea
(dashed line) and estimated paths of the Huanghe and Changing Rivers
(dotted lines) are also illustrated to show the paleo-positions of the river
mouths during glacial stages. Dotted line in the Japan Sea represents
bathymetric contour of -2000m.

Japanese Islands and mostly flows out through the Tsugaru Strait to the
Pacific Ocean. The upper 50 m of the TWC is characterized with
slightly lower salinity of 33.7 to 34.0 permil and low phosphate
concentrations of approximately 0.1 μ mol/kg, whereas the TWC below
50 m has higher salinity and phosphate concentrations in excess of 34.4
permil and 0.5 μ mol/kg, respectively [4]. The coastal water of Eastern
China Sea (ECS) and Yellow Sea is characterized with lower salinity of
31 to 32 permil and higher phosphate concentrations of 0.5 μ mol/kg,
respectively [3].

The deep water called the Japan Sea Proper Water (JSPW) is formed in
the northern part of the sea during winter either through cooling of the
surface water with slightly higher salinity inherited from the TWC or
through formation of sea ice [10,11]. Due to the high ventilation rate of
several hundreds' years, deep water of the Japan Sea is characterized with
extremely high level of dissolved oxygen (>210 mmol/kg), low level of
phosphate (<2 μ mol/kg), and low and nearly constant temperature of
approximately 0°C [2]. It is expected that a glacio-eustatic sea-level

change had caused significant in the oceanographic conditions of the Japan Sea during the Plio-Pleistocene.

Background
Late Pliocene to Holocene hemipelagic sediments recovered from the basinal sites by ODP Legs 127 and 128 are characterized with distinctly dark and light color-banded clay that is slightly to moderately diatomaceous and occasionally calcareous (Fig. 2). These dark and light layers are correlatable basinwide and the their deposition was synchronous judging from the their parallel relationship with the marker tephra layers. Dark layers are olive black and generally laminated, whereas light layers are light gray and homogeneous to bioturbated. The darkness of the layers is primarily reflecting organic carbon content that ranges from 1 to 7 % in dark layers and 0.1 to 1 % in light layers [14]. Diatom assemblages in these sedimentary sequence show the changes in the millennial to sub-Milankovitch scale with less abundance during glacial periods and abundant during interglacial periods [9].

The warm plus coastal-water diatom abundance suggests the dominant modulation of the inflow through the Tsushima Strait by eustatic sea-level changes. Superimposed on this trend are millenial-scale. Oscillations in the diatom abundance. Correlation of the diatom abundance with the organic carbon show that the deposition of dark layers generally coincides with the stronger inflow peaks, but the peaks lead dark layer deposition by 1 to 2 kyr in several cased. Close examination of such ages reveals that the warm-water diatoms dominate these peaks, and deposition of such dark layers started only after the ECSCW inflow became dominant.

Present ECSCW is highly enriched in nutrients as compared with the upper TWC due to intensive regeneration of the nutrients at the shelf floor. Consequently, it is likely that stronger inflow of the ECSCW supplied more nutrients and caused higher productivity in the Japan Sea. In addition, its stronger inflow would also have strengthened the density stratification within the sea due to its slightly lower salinity. The combination of these two effects would have resulted in development of anoxic to euxinic bottom-water and deposition of dark layers.

Higher ECSCW influx during inter mediate sea-levels compared to low sea-levels resulted in higher productivity and stronger reducing condition of the bottom-water during deposition of the dark layers. On the other hand, the nutrients supply was reduced and density stratification was disrupted during the intervals of smaller influx of the ECSCW leading to deposition of light layers with rare warm-water plus coastal-water diatoms [15].

MATERIALS AND METHODS

The ODP Site 797 was selected as locating the critical boundary of water masses in the Japan Sea [16]. At Site 797, a core gap of approximately 35 cm thick was recognized between Cores 797B-1H and 797B-2H [14], which was supplemented by Core 797A-1H (Table 1). The Plio-Pleistocene sequence of 83.90 cmbs (corrected meter below surface) of clay and silty clay which are occasionally biosiliceous and/or biocalcareous was recovered from Hole 797B at 38.616°N, 134.536°E; water-depth 2862.2 m with supplemented one sample of 797A-1H-3, 18-20 cm (Table 1). They show centimeter to decimeter-scale alternation of the dark and light layers (Fig. 2) which are correlatable within the Japan Sea [14]. Such marker tephra layers as Aira-Tanzawa (AT), Aso-4, Ata-Th, Aso-1, Beagdusan-Oga (B-Og), and Baegdusan are identified at 2.43, 8.80, 24.40, 25.50, 44.70, and 58.00 cmbs, respectively [13] (Table 1).

274 samples obtained with the average sampling interval of approximately 0.31 cm from 83.90 cmbs in the Plio-Pleistocene sequence of ODP Site 797 (Table 1, Fig. 2). Ages of these samples were constrained by six well-dated tephras and magneto stratigraphy. Depth-age profile, which is assumed o ka age for the core-top and constant linear sedimentation rate between these ten datum levels, indicates the sampling interval averages 6,567 year (Table 1).

TREATMENT OF MATERIALS AND DIATOM ANALYSIS

Quantitative diatom slides were prepared [8]. Several samples had less diatoms than 200 specimens in total. The numbers of diatom specimens per 1 g of the dried sample were calculated. Diatom abundance is regarded as reflecting combined effect of productivity and preservation. The increase in abundance of a low-salinity coastal-water diatom *Paralia sulcata* is regarded as an indicator of the ECSCW during sea-level intermediate stands [5,6,7,16], and the dominance of the warm-water diatoms is regarded as an indicator of the TWC during high stands (Table 2). Since both the ECSCW and TWC make the inflowing waters through the Tsushima Strait, the warm-plus coastal-water diatom abundance is considered as indicator of the inflow strength. Superimposed on this trend are millenial-scale oscillation in the diatom abundance (Fig. 2).

DIATOM ASSEMBLAGE

Using three times of the distinct changes in the abundance of diatom, *P. sulcata*, and inflow, the Plio-Pleistocene diatom record can be divided

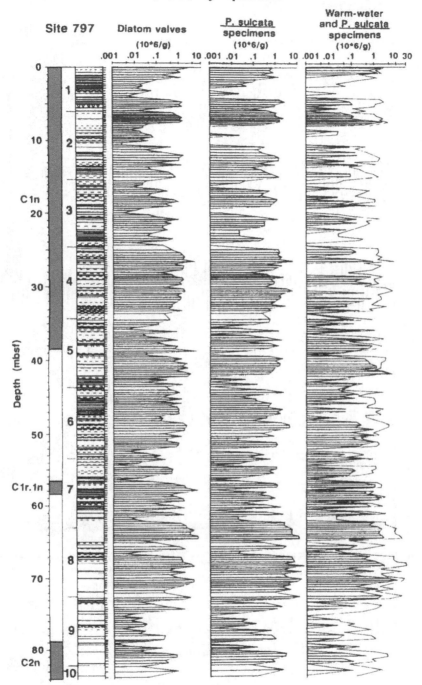

Figure 2. Stratigraphic variations of diatom abundance (per g) and selected diatom species in the Plio-Pleistocene since 2,000 kyr ago at ODP Site 797. Black=dark layer, white=light layer, dashed lines=faintly dark layer, and v=ash layer.

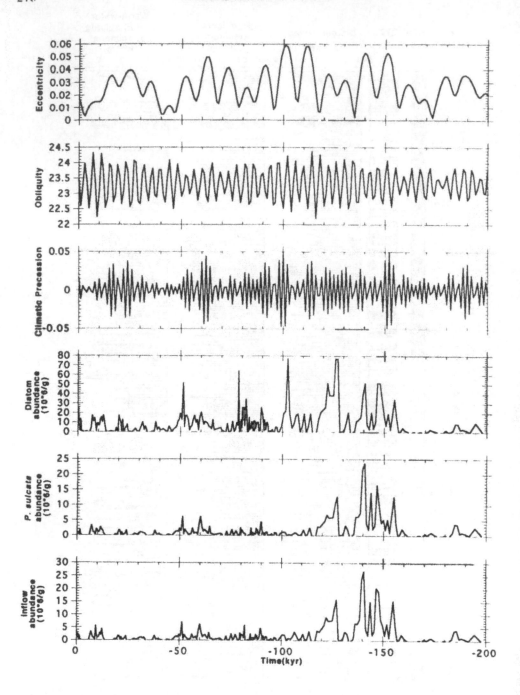

Figure 3. Comparison between stratigraphic variations of three kinds of proxy indicators of surface-water condition and the Earth's orbital parameters with the insolation at 37°N [1] during the last 2,000 kyr at ODP Site 797.

into four sections: (1) the latest Pliocene (2.0-1.6 Ma), (2) the early Pleistocene (1.6-1.1 Ma), (3) the middle Pleistocene (1.1-0.5 Ma), and the late Pleistocene (0.5-0 Ma) (Fig. 3). Throughout the latest Pliocene of the diatom record the amplitude of abundance of diatom, *P. sulcata*, and inflow components is low, with small fluctuation. At about 1.6 Ma an increase in amplitude of these components occur and these phases are sympathized with each other. This trend continues to about 1.1 Ma when a marked decrease in amplitude occurred. These components gradually decrease from the interval between 1.1 and 0.5 Ma to the interval younger than 0.5 Ma. In contrast, the amplitude record for the variations in the Earth's orbital parameters, which are proposed as external forcing function of climatic change does not match visually, peak to peak, amplitude variations in these components.

SPECTRAL ANALYSIS OF DIATOM ABUNDANCE

The irregular time interval in native time series of these three sections (Figs. 4-6(A) left-upper) was normalized by equal sampling time intervals as the modified time series (Figs. 4-6(A) left-lower). The spectral estimation (Figs. 4-6(A) middle-upper and (B) left) and the fitting test of the most suitable curve for the modified time series (Figs. 4-6(B) right) were analyzed on the system soft-ware for time series data analysis MEMCALC 1000 (Suwa Trust Co., Ltd.). This soft-ware is based on both the Maximum Entropy Method (MEM) and the nonlinear Least Squares Fitting (LSF) method. In order to precisely estimate MEM-power spectral density (MEM-PSD), suitable criteria are required for determining the optimum value M (a minimum-phase of the prediction-error filter). In this study, M is situated between the characteristic correlation time (CCT) and number of data (N)/2, and is determined by the minimization of three conventional criteria: the final prediction error (fpe) criterion, Akaike's information theoretical criterion (aic), and the auto-regressive transfer function criterion (cat). These time series are not modified by any tuning or stacking.

The early Pleistocene (1.6-1.1 Ma)
An arrow indicates M at 30 between CCT (=30) and N (number of data= 165)/2 (=82.5) adopted for the MEM-PSD analysis (Fig. 3(A) right). A log-log plot of the MEM-PSD shows a decreasing trend of 1/f3-like dependence which is considered to be strongly bound physically (Fig. 3(A) middle-upper). The MEM-PSD (linear scale) consist of one predominant 203.7 kyr period peak and three minor periods: 70.2, 40.4, and 28.6 kyr (Fig. 3(B) left). The optimum LSF curve (underlying variation) was calculated using these four periods (Fig. 3(B) right).

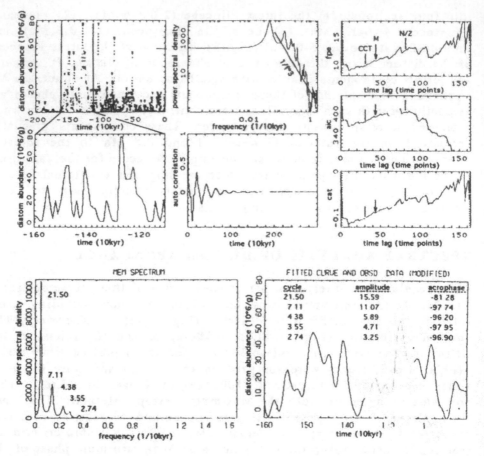

Figure 4. Diatom abundance in the early Pleistocene (1.6-1.1 Ma) at ODP Site 797. (A) left-upper: Native time series; left-lower: Modified time series; middle-upper: Log-log representation of the MEM-PSD; middle-lower: Auto correlation; right: Information criteria (fpe=final prediction error, aic=Akaike's information criterion, cat=criterion autoregressive transfer function). Bars indicate a characteristic correlation time, CCT at 32 and N (number of samples)/2 at 82.5, and an arrow shows a minimum-phase of prediction error filter (time rag), M at 44. (B) left: Linear-linear representation of the MEM-PSD; right: Comparison the optimum LSF curve with the original data.

The Middle Pleistocene (1.1-0.5 Ma)

The spectrum was analyzed based on 19 of CCT, 99 of N/2, and 44 of M, show 1/f2-dependence, a brown noise which is a peculiar to the so-called Brownian motion, of the MEM-PSD (log-log scale) (Fig. 4 (A) middle- upper). It is composed of predominant 224.0 and 112.2 kyr, and five minor: 57.3, 42.8, 31.0, 25.2, and 18.3 kyr periods (Figs. 4 (B) left).

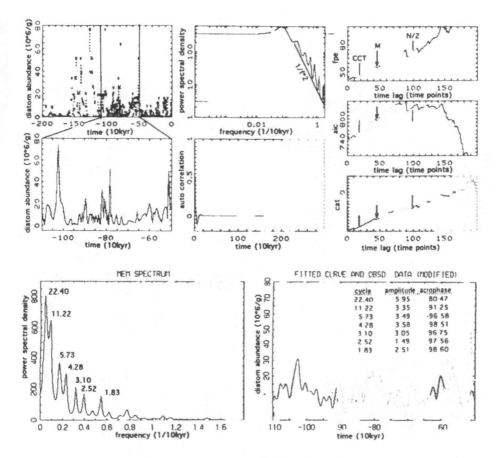

Figure 5. Diatom abundance in the middle Pleistocene (1.1-0.5 Ma) at ODP Site 797. (A) left-upper: Native time series; left-lower: Modified time series; middle-upper: Log-log representation of the MEM-PSD; middle-lower: Auto correlation; right: Information criteria (fpe=final prediction error, aic=Akaike's information criterion, cat=criterion autoregressive transfer function). Bars indicate a characteristic correlation time, CCT at 19 and N (number of samples)/2 at 99, and an arrow shows a minimum-phase of prediction error filter (time rag), M at 44. (B) left: Linear-linear representation of the MEM-PSD; right: Comparison the optimum LSF curve with the original data.

The Late Pleistocene (0.5-0 Ma)

The MEM-PSD based 23 of CCT, 82 of N/2, and 27 of M show $1/f2$-dependence (Fig. 5 (A)). There is a very strong peak at 105.3 kyr period and this is accompanied with three minor 40.0, 18.3, and 25.8 kyr components (Fig. 5 (B) left). The optimum LSF curve is in fairly good agreement with the original time series data on the whole, except for the deviations around 400 ka (Fig. 5 (B) right).

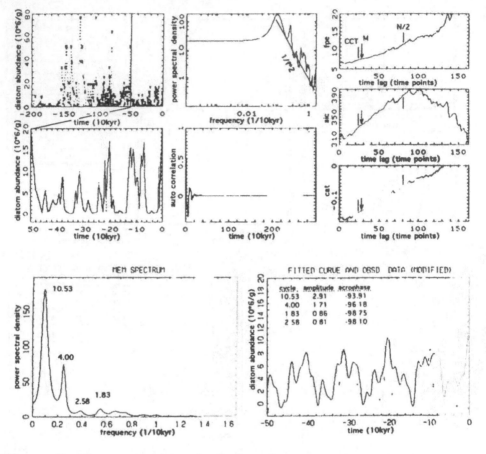

Figure 6. Diatom abundance in the late Pleistocene (0.5-0 Ma) at ODP Site 797. (A) left-upper: Native time series; left-lower: Modified time series; middle-upper: Log-log representation of the MEM-PSD; middle-lower: Auto correlation; right: Information criteria (fpe=final prediction error, aic=Akaike's information criterion, cat=criterion autoregressive transfer function). Bars indicate a characteristic correlation time, CCT at 23 and N (number of samples)/2 at 82, and an arrow shows a minimum-phase of prediction error filter (time rag), M at 27. (B) left: Linear-linear representation of the MEM-PSD; right: Comparison the optimum LSF curve with the original data.

DISCUSSION AND CONCLUSIONS

The Pliocene-Pleistocene boundary (1.6 Ma) may have been a slight decrease in temperatures and /or an increase in precipitation within the transition zone [5,12]. The early Pleistocene (1.6-1.1 Ma) would have resulted in fluctuations of both temperature and precipitation but variations in temperature were less extreme than those of precipitation. Diatom abundance is in remarkably decreased around 1.35 Ma in the early Pleistocene (Fig. 3). It was possibly affected by a slight southward shift of the subarctic water mass occurred at 1.3 Ma. The central subarctic region experienced a marked drop in temperature, with increased precipitation and stratification at this time.

About 1.1 Ma a marked decrease occurred in surface water temperatures in the northwest Pacific. Transport by the Oyashio Current increased, resulting in strong mixing and high production along the western boundary. Around 0.8 Ma diatom abundance of the Japan Sea increased. At this time the subarctic front expanded southward in the central North Pacific and have caused a steepening of salinity gradient. As a result, cold upwelling waters became the marine source of subarctic surface water [12].

Since 0.5 ma the lithologic and diatom record in the Japan Sea and subarctic sites has shown a strong 100- kyr periodicity, with two stable climatic extremes, glacial and interglacial condition [5]. The trend shows the lowering sea level by glaciation more severe than in earlier times.

Predominant cycles of 112.2 kyr in the middle part and 105.3 kyr in the late part of the Pleistocene correspond to 95-124 kyr cycle of the orbital eccentricity among the Earth's orbital parameters (Fig. 3). This first-order time series in the diatom abundance of the Japan Sea started at 1.1 Ma of the middle Pleistocene and is characterized by the glacial-interglacial cycle of decreasing frequencies with time. The cycles of 203.7 kyr in the early part and 224.0 kyr in the middle part are twice the eccentricity (95-124 kyr cycle). Diatom abundance increases when nutrient rich middle-deep sea-water is provided to the surface as upwelling during transgressive and interglacial periods. During regressive and glacial periods, diatom abundance drastically decreased when the lowering of sea-level causes shallowing of the strait and retarding of the inflow of middle-deep sea-water into the Japan Sea [15].

The 2nd-order time series which are composed of the ranging from 40.0 kyr to 42.8 kyr cycle correspond to the obliquity (tilt) band (41-54 kyr) variations (Fig. 3). The periods of 28.6 kyr in the early part and 29 kyr period in nonlinear interactions of the 100 kyr period of eccentricity and the 41 kyr period of tilt. The 18.3 kyr cycle in the middle-late part corresponds to the precession band (19-23 kyr) variations. There are two kinds of periods from 70.2 kyr in the early Pleistocene to 57.3 kyr in the middle Pleistocene and from 25.2 kyr in the middle to 25.8 kyr in the late which are not present in the primary Milankovitch cycles.

REFERENCES

1. A. Berger and M. F. Loutre. Insolation values for the climate of the last 10 million of years, *Quaternary Sciences Review* 10, 297-317 (1991).
2. T. Gamo, Y. Nozaki H. Sakai, T. Nakai and H. Tsubota. Spatial and temporal variations of water characteristics in the Japan Sea, *J. Mar. Res* 44, 781-793 (1986).
3. G. H. hong, S. H. Kim, C. S. Chung, D. J. Kang, J. Choi, T. S. Lee and J. Y. Chung. Nutrients and biogeochemical provinces in the Yellow Sea. In: *Global Fluxes of Carbon and Its Related Substances on the Coastal Sea-Ocean-Atmosphere System*. S. Tsunogai et al. (Eds.). pp. 114-128. M&J International, Yokohama (1995).
4. Y. Horibe (Ed.). *Preliminary Report of the Hakuho Maru Cruise KH-77-3 (Pegasus Expedition)*. Ocean Research Institute, University of Tokyo (1981).
5. I. Koizumi. Late Neogene Paleoceanography on the western north Pacific, *Init. Rep. DSDP* 86, 429-438 (1985).
6. I. Koizumi. Pulses of Tsushima Current during the Holocene, *the Quaternary Research*. 26, 13-25 (1987).
7. I. Koizumi. Holocene pulses of diatom growths in the warm Tsushima Current in the Japan Sea, *Diatom Res*. 4, 55-68 (1989).
8. I. Koizumi. Diatom biostratigraphy of the Japan Sea: Leg 127. In: *Proc. ODP, Sci. Res., 127/128, Pt. 1*. K. A. Pisciotto et al. (Eds.). pp. 249-288. College Station, TX (Ocean Drilling Program). (1992a).
9. I. Koizumi. Biostratigraphy and paleoceanography of the Japan Sea based on diatoms: ODP Leg 127. In: *Pacific Neogene-Environment, Evolution, and Events*. R. Tsuchi and J. C. Ingle (Eds.). pp. 15-24. University of Tokyo Press, Tokyo. (1992b).
10. S. Martin, E. Munoz and R. Drucker. The effect of severe storms on the ice cover of the northern Tatarskiy Strait, *J. Geophys. Res.* 97, 17753-17764 (1992).
11. K. Nishiyama, S. Kawae and H. Sasaki. The Japan Sea Proper Water and the Japan Sea Warm Eddy (in Japanese with English abstract), *Bull. Kobe Marine Observatory*. 209, 1-10 (1990).
12. C. Sancetta and S. M. Silvestri. Pliocene-Pleistocene evolution of the North Pacific ocean-atmosphere system, interpreted from fossil diatoms, *Paleoceanography*. 1, 163-180 (1986).
13. R. Tada. The solving for the aridity of central Asia and the strength of prevailing winds during the Quaternary, *The report of scientific results of a Grant-in-Aid for Scientific Research from the Ministry of Education, Science and Ministry of Education*. pp. 105. Univ. of Tokyo, Tokyo (1996).
14. R. Tada, I. Koizumi, A. Cramp and A. Rahman. Correlation of dark and light layers, and the origin of their cyclicity in the Quaternary sediments from the Japan Sea. In: *Proc. ODP, Sci. Res., 127/128, Pt. 1*. K. A. Pisciotto et al. (Eds.). pp. 577-601. College Station, TX (Ocean Drilling Program). (1992).
15. R. Tada T. Irino and I. Koizumi. Possible Dansgaard-Oeschger oscillation signal recorded in the Japan Sea sediments. In: *Global Fluxes*

of Carbon and Its Related Substances in the Coastal Sea-Ocean-Atmosphere System. S. Tsunogai et al. (Eds.). pp. 517-522. M&J International, Yokohama (1995).

16. Y. Tanimura. Late Quaternary diatoms and Paleoceanography of the Sea of Japan (in Japanese with English abstract). *The Quaternary Research.* 20, 231-242 (1981).

Table 1. Ages and abundances of proxy indicators for the surface-water condition since 2,000,000 years ago at ODP Site 797. (cc=core-catchar)

Hole	Core-section interval (cm)		Depth (cmbsf)	Time (10ka)	Diatom abundance (10^{*6}/g)	P.sulcata speciments (10^{*6}/g)	Warm-water speciments (10^{*6}/g)	Inflow abundance (10^{*6}/g)
B	1H-1,	13-15	0.13	0.14	25.71	0.77	2.19	2.96
		45-47	0.44	0.48	7.71	0.58	0.58	1.16
		74-76	0.71	0.77	11.87	0.89	1.42	2.31
		101-103	0.97	1.05	1.37	0.21	0.06	0.28
		134-136	1.28	1.39	12.86	1.67	0.00	1.67
	2,	13-15	1.56	1.69	1.21	0.15	0.00	0.15
		44-46	1.86	2.02	0.59	0.10	0.00	0.10
		74-76	2.14	2.32	0.18	0.03	0.00	0.03
		84-86	2.24	2.43	AT			
		104-106	2.43	2.84	0.26	0.05	0.00	0.05
		138-140	2.75	3.52	0.11	0.00	0.00	0.00
	3,	13-15	2.99	4.04	0.69	0.00	0.00	0.00
		44-46	3.28	4.66	0.18	0.03	0.00	0.03
		74-76	3.57	5.28	0.05	0.00	0.00	0.00
		104-106	3.85	5.88	0.15	0.00	0.00	0.00
		134-136	4.14	6.51	4.29	1.70	0.03	1.72
	4,	13-15	4.41	7.08	17.14	3.17	0.09	3.26
		44-46	4.71	7.73	11.02	1.05	0.11	1.16
		71-73	4.97	8.29	14.69	1.03	0.00	1.03
		98	5.21	8.80	Aso-4			
		104-106	5.28	8.91	1.59	0.07	0.03	0.10
		109-111	5.33	8.99	0.96	0.51	0.00	0.51
		144-146	5.56	9.35	7.70	2.08	3.55	5.63
A	1H-3,	18-20	5.79	9.70	9.40	1.78	0.09	1.87
B	2H-1,	14-16	6.04	10.09	6.43	1.00	0.03	1.03
		44-46	6.33	10.55	7.71	1.04	0.08	1.12
		67-69	6.56	10.90	12.86	1.22	0.13	1.35
		102-104	6.89	11.42	15.43	2.70	0.08	2.78
		134-136	7.20	11.90	10.29	1.75	0.36	2.11
	2,	14-16	7.49	12.35	17.14	1.71	2.66	4.37
		44-46	7.78	12.81	12.86	0.77	0.77	1.54
		74-76	8.07	13.26	5.14	0.28	0.31	0.59
		102-104	8.34	13.68	0.28	0.00	0.00	0.00
		134-136	8.65	14.16	0.26	0.00	0.00	0.00
	3,	14-16	8.94	14.61	0.05	0.00	0.00	0.00
		44-46	9.23	15.06	0.18	0.00	0.03	0.03
		74-76	9.52	15.52	0.23	0.00	0.00	0.00
		102-104	9.79	15.94	0.62	0.03	0.00	0.03
		134-136	10.09	16.41	0.77	0.00	0.00	0.00
	4,	14-16	10.39	16.87	0.13	0.00	0.00	0.00
		44-46	10.68	17.32	0.44	0.00	0.00	0.00
		74-76	10.97	17.78	0.03	0.00	0.00	0.00
		102-104	11.24	18.20	0.03	0.00	0.00	0.00
		134-136	11.55	18.68	4.47	0.33	0.08	0.41
	5,	16-18	11.86	19.16	1.05	0.15	0.00	0.15
		44-46	12.13	19.58	1.85	0.28	0.15	0.44
		74-76	12.42	20.04	17.14	0.86	0.17	1.03
		102-104	12.69	20.46	12.86	1.74	0.00	1.74
		134-136	12.99	20.92	10.29	1.49	0.05	1.54

Hole	Core-section interval (cm)		Depth (cmbsf)	Time (10ka)	Diatom abundance (10*6/g)	P.sulcata speciments (10*6/g)	Warm-water speciments (10*6/g)	Inflow abundance (10*6/g)
B	2H-6,	14-16	13.28	21.38	0.44	0.13	0.00	0.13
		44-46	13.57	21.83	12.86	0.96	0.19	1.16
		74-76	13.86	22.28	0.39	0.08	0.00	0.08
		104-106	14.15	22.73	0.58	0.24	0.00	0.24
		134-136	14.44	23.18	3.91	0.44	0.00	0.44
	7,	14-16	14.73	23.64	5.14	0.77	0.08	0.85
		44-46	15.02	24.09	7.71	1.58	0.08	1.66
	cc,	4	15.22	24.40	Ata-Th			
	3H-1,	3-5	15.28	24.55	0.41	0.04	0.00	0.04
		14-16	15.55	25.23	0.28	0.03	0.00	0.03
		27	15.66	25.50	Aso-1			
		44-46	15.84	25.94	0.23	0.00	0.00	0.00
		74-76	16.13	26.64	0.08	0.00	0.00	0.00
		103-105	16.41	27.32	1.47	0.05	0.03	0.08
		135-137	16.72	28.07	3.50	0.54	0.00	0.54
	2,	14-16	17.00	28.75	0.77	0.31	0.00	0.31
		44-46	17.29	29.45	0.46	0.23	0.00	0.23
		74-76	17.58	30.15	1.29	0.52	0.03	0.54
		103-105	17.86	30.83	5.93	1.04	0.00	1.04
		135-137	18.17	31.58	10.29	1.03	0.10	1.13
	3,	14-16	18.45	32.26	2.40	0.71	0.02	0.73
		40-42	18.71	32.89	5.71	0.63	0.43	1.06
		74-76	19.04	33.69	0.26	0.00	0.00	0.00
		103-105	19.32	34.37	0.10	0.00	0.00	0.00
		135-137	19.63	35.12	0.03	0.00	0.00	0.00
	4,	14-16	19.91	35.80	0.28	0.03	0.00	0.03
		44-46	20.20	36.51	0.44	0.03	0.00	0.03
		74-76	20.49	37.21	0.98	0.13	0.03	0.15
		103-105	20.77	37.89	10.29	1.54	0.15	1.70
		135-137	21.08	38.64	2.08	0.36	0.03	0.39
	5,	14-16	21.36	39.32	4.82	0.39	0.17	0.55
		44-46	21.65	40.02	2.11	0.39	0.00	0.39
		73-75	21.93	40.70	2.08	0.21	0.00	0.21
		103-105	22.22	41.40	3.37	0.03	0.05	0.08
	6,	14-16	22.82	42.86	0.64	0.03	0.00	0.03
		44-46	23.11	43.56	0.39	0.03	0.00	0.03
		73-75	23.39	44.24	5.71	0.31	0.00	0.31
		94	23.58	44.70	B-Og			
		103-105	23.68	44.93	2.86	0.21	0.00	0.21
		135-137	23.99	45.63	0.10	0.00	0.00	0.00
	7,	14-16	24.27	46.26	1.89	0.00	0.00	0.00
	4H-1,	14-17	25.04	48.00	5.71	1.26	0.09	1.34
		44-46	25.32	48.64	11.02	1.43	0.17	1.60
		74-76	25.60	49.27	12.86	1.16	0.06	1.22
		102-104	25.86	49.86	19.29	2.31	0.00	2.31
		135-137	26.16	50.54	12.86	1.48	0.13	1.61
	2,	20-22	26.49	51.28	51.43	6.17	0.77	6.94
		44-46	26.71	51.78	5.14	0.36	0.00	0.36
		74-76	26.99	52.41	17.14	2.06	0.00	2.06
		103-105	27.26	53.02	17.14	1.29	0.00	1.29
		135-137	27.56	53.70	12.86	0.71	0.00	0.71
	3,	14-16	27.83	54.31	1.67	0.07	0.00	0.07

Hole	Core-section interval (cm)		Depth (cmbsf)	Time (10ka)	Diatom abundance (10 *6/g)	P.sulcata speciments (10 *6/g)	Warm-water speciments (10 *6/g)	Inflow abundance (10 *6/g)
B	4H-3,	44-46	28.11	54.95	5.41	0.46	0.00	0.46
		74-76	28.39	55.58	6.43	0.58	0.00	0.58
		102-104	28.65	56.17	12.86	2.38	0.00	2.38
		135-137	28.95	56.85	17.14	0.60	0.26	0.86
	4,	13-15	29.21	57.43	9.64	0.96	0.10	1.06
		41	29.46	58.00		Baegdusan		
		44-46	29.50	58.09	12.86	0.90	0.64	1.54
		74-76	29.78	58.72	1.98	0.33	0.00	0.33
		102-104	30.04	59.31	15.43	3.63	0.00	3.63
		135-137	30.35	60.01	20.57	6.17	0.00	6.17
	5,	14-16	30.62	60.62	12.89	2.84	0.32	3.16
		43-45	30.89	61.24	7.71	1.31	0.00	1.31
		71-73	31.15	61.82	11.02	1.93	0.00	1.93
		102-104	31.43	62.46	10.29	1.54	0.62	2.16
		135-137	31.74	63.16	8.57	0.90	0.21	1.11
	6,	14-16	32.01	63.77	5.14	0.59	0.05	0.64
		44-46	32.29	64.40	5.14	2.88	0.05	2.93
		73-75	32.56	65.01	2.96	0.54	0.00	0.54
		102-104	32.83	65.62	12.89	0.58	0.19	0.77
		135-137	33.13	66.30	2.06	0.26	0.00	0.26
	5H-1,	14-16	34.54	69.49	4.53	0.62	0.10	0.72
		44-46	34.83	70.15	1.13	0.49	0.00	0.49
		74-76	35.12	70.81	1.16	0.08	0.03	0.10
		103-105	35.40	71.44	0.10	0.00	0.00	0.00
		132-134	35.68	72.07	0.05	0.03	0.00	0.03
	2,	17-19	36.02	72.84	8.57	1.33	0.26	1.59
		44-46	36.28	73.43	0.23	0.00	0.00	0.00
		74-76	36.57	74.09	0.32	0.06	0.00	0.06
		103-105	36.85	74.72	1.44	0.41	0.00	0.41
		132-134	37.13	75.35	11.02	1.93	0.55	2.48
	3,	14-16	37.44	76.05	0.93	0.05	0.03	0.08
		44-46	37.73	76.71	2.01	0.62	0.05	0.67
		74-76	38.02	77.37	9.64	1.45	0.72	2.17
		102-104	38.29	77.98	7.01	0.49	0.49	0.98
			38.30	78.00		B/M		
		132-134	38.58	78.31	64.00	0.05	0.10	0.15
	4,	14-16	38.88	78.65	0.39	0.00	0.03	0.03
		42-45	39.16	78.96	0.31	0.00	0.00	0.00
		74-76	39.46	79.29	3.40	0.54	0.03	0.57
		103-105	39.74	79.60	12.86	2.51	0.32	2.83
		131-133	40.01	79.90	4.45	0.91	0.08	0.99
	5,	11-13	40.30	80.22	25.71	1.80	0.51	2.31
		43-45	40.61	80.57	12.86	1.42	0.58	1.99
		74-76	40.91	80.90	25.71	0.90	1.16	2.06
		103-105	41.19	81.21	6.43	0.80	0.16	0.97
		131-133	41.46	81.51	17.14	0.69	2.14	2.83
	6,	14-16	41.78	81.87	34.29	0.51	5.32	5.83
		44-46	42.07	82.19	25.71	0.00	0.13	0.13
		81-83	42.42	82.58	0.32	0.00	0.00	0.00
		102-104	42.63	82.82	2.86	0.04	0.03	0.07
		131-133	42.91	83.13	0.26	0.00	0.00	0.00
	7,	14-16	43.22	83.47	15.43	0.15	0.15	0.31

Hole	Core-section interval (cm)		Depth (cmbsf)	Time (10ka)	Diatom abundance (10*6/g)	P.sulcata speciments (10*6/g)	Warm-water speciments (10*6/g)	Inflow abundance (10*6/g)
B	5H-7,	43-45	43.50	83.78	2.57	0.00	0.08	0.08
	6H-1,	13-15	44.03	84.37	4.24	0.30	0.03	0.32
		43-45	44.32	84.70	11.02	2.42	0.06	2.48
		74-76	44.62	85.03	4.65	0.28	0.18	0.46
		102-104	44.89	85.33	1.47	0.03	0.00	0.03
		132-134	45.18	85.65	11.43	0.91	0.11	1.03
	2,	13-15	45.47	85.98	1.13	0.13	0.00	0.13
		43-45	45.76	86.30	11.43	0.63	0.46	1.09
		74-76	46.06	86.63	3.57	0.49	0.00	0.49
		102-104	46.33	86.93	7.91	0.44	0.00	0.44
		132-134	46.62	87.25	10.29	2.32	0.05	2.37
	3,	13-15	46.91	87.58	9.64	0.96	0.05	1.01
		43-45	47.20	87.90	10.29	1.80	0.00	1.80
		74-76	47.50	88.23	1.22	0.39	0.00	0.39
		102-104	47.77	88.53	1.69	0.39	0.02	0.41
		132-134	48.06	88.86	1.44	0.84	0.00	0.84
	4,	13-15	48.35	89.18	14.69	1.98	0.15	2.13
		43-45	48.64	89.50	25.71	4.37	0.26	4.63
		74-76	48.94	89.84	20.57	4.32	0.10	4.42
		102-104	49.21	90.14	19.28	1.74	0.77	2.51
	5,	13-15	49.79	90.78	1.44	0.24	0.06	0.30
		43-45	50.08	91.10	12.86	0.84	0.84	1.67
		74-76	50.38	91.44	2.94	0.17	0.09	0.26
		102-104	50.65	91.74	2.10	0.15	0.17	0.32
		132-134	50.93	92.05	9.64	0.96	0.19	1.16
	6,	13-15	51.23	92.38	10.29	1.18	0.31	1.49
		43-45	51.52	92.70	7.35	1.07	0.18	1.25
		74-76	51.82	93.04	0.34	0.00	0.00	0.00
		102-104	52.09	93.34	0.04	0.00	0.00	0.00
		132-134	52.37	93.65	0.28	0.09	0.00	0.09
	7,	13-15	52.67	93.98	1.29	0.28	0.04	0.32
		43-45	52.96	94.31	0.60	0.04	0.02	0.06
	7H-1,	12-14	53.52	94.93	2.61	0.62	0.11	0.73
		43-45	53.82	95.26	0.26	0.07	0.02	0.09
		70-72	54.08	95.55	0.11	0.02	0.00	0.02
		102-104	54.38	95.89	0.00	0.00	0.00	0.00
		132-134	54.67	96.21	2.96	0.97	0.00	0.97
	2,	12-14	54.95	96.52	5.51	1.05	0.08	1.13
		43-45	55.25	96.85	5.14	0.57	0.13	0.69
		74-76	55.54	97.18	5.14	0.85	0.08	0.93
		102-104	55.81	97.48	0.64	0.17	0.06	0.24
		132-134	56.09	97.79	0.21	0.02	0.00	0.02
	3,	12-14	56.38	98.11	0.00	0.00	0.00	0.00
			56.64	98.40	Top of Jaramillo			
		43-45	56.67	98.51	0.64	0.02	0.00	0.02
		74-76	56.97	99.57	7.01	0.91	0.35	1.26
		102-104	57.24	100.53	19.29	1.16	1.06	2.22
		132-134	57.52	101.52	22.04	1.10	0.99	2.09
	4,	10-12	57.79	102.48	77.14	0.00	0.00	0.00
		41-43	58.08	103.51	19.29	0.39	0.39	0.77
		74-76	58.40	104.64	4.29	0.19	0.34	0.54
		102-104	58.66	105.57	7.91	0.55	0.79	1.35

Hole	Core-section interval (cm)		Depth (cmbsf)	Time (10ka)	Diatom abundance (10 *6/g)	P.sulcata speciments (10 *6/g)	Warm-water speciments (10 *6/g)	Inflow abundance (10 *6/g)
B	7H-4,	132-134	58.95	106.59	17.14	0.94	2.31	3.26
	5,	12-14	59.24	107.62	19.29	0.48	1.06	1.54
		43-45	59.53	108.65	0.41	0.02	0.00	0.02
		74-76	59.83	109.72	5.51	0.41	0.41	0.83
		102-104	60.09	110.64	19.29	1.83	0.58	2.41
		132-134	60.38	111.67	0.06	0.00	0.00	0.00
	6,	12-14	60.66	112.66	5.51	0.44	0.08	0.52
		43-45	60.96	113.73	19.29	2.51	0.97	3.47
		74-76	61.25	114.75	0.06	0.02	0.00	0.02
		102-104	61.52	115.71	0.11	0.02	0.00	0.02
		132-134	61.81	116.74	0.30	0.06	0.00	0.06
	7,	12-14	62.09	117.73	12.86	2.89	0.90	3.79
		43-45	62.39	118.80	20.57	2.78	0.51	3.29
	8H-1,	13-15	63.04	121.10	25.71	4.63	1.29	5.91
		43-45	63.32	122.10	51.43	6.69	2.31	9.00
		74-76	63.62	123.16	38.57	5.79	1.74	7.52
		132-134	64.18	125.15	38.57	5.98	1.74	7.71
	2,	13-15	64.48	126.21	77.14	10.03	2.31	12.34
		43-45	64.77	127.24	77.14	12.73	3.09	15.81
		74-76	65.07	128.31	0.17	0.06	0.00	0.06
		132-134	65.63	130.29	0.99	0.28	0.00	0.28
	3,	13-15	65.93	131.36	9.35	3.27	0.14	3.41
		43-45	66.22	132.39	20.57	2.06	0.72	2.78
		74-76	66.52	133.45	0.24	0.02	0.00	0.02
		102-104	66.79	134.41	0.04	0.02	0.00	0.02
		132-134	67.08	135.44	0.94	0.58	0.06	0.64
	4,	10-12	67.35	136.39	11.43	6.12	0.51	6.63
		41-43	67.65	137.46	12.86	6.11	0.26	6.37
		74-76	67.97	138.59	15.43	7.18	0.93	8.10
		102-104	68.24	139.55	38.57	21.60	1.74	23.34
		132-134	68.53	140.58	51.43	23.66	3.09	26.74
	5,	10-12	68.80	141.54	7.71	3.97	0.27	4.24
		43-45	69.12	142.67	3.09	1.93	0.04	1.97
		74-76	69.42	143.74	20.57	14.09	0.82	14.91
		102-104	69.69	144.69	3.73	2.04	0.04	2.08
		132-134	69.98	145.72	10.29	3.34	0.41	3.76
	6,	12-14	70.23	146.61	51.43	16.46	3.86	20.32
		43-45	70.57	147.82	51.43	9.26	9.52	18.77
		74-76	70.86	148.85	34.29	4.12	3.94	8.06
		102-104	71.14	149.84	17.14	5.14	0.51	5.66
		129-131	71.40	150.76	6.43	1.74	0.13	1.87
	7,	12-14	71.70	151.83	19.29	5.11	1.35	6.46
		42-44	72.00	152.89	2.46	0.81	0.17	0.98
	9H-1,	14-46	72.54	154.81	34.29	12.69	2.57	15.26
		43-45	72.82	155.80	15.43	4.09	1.62	5.71
		74-76	73.12	156.86	0.15	0.00	0.00	0.00
		103-105	73.39	157.82	0.73	0.41	0.00	0.41
		132-134	73.67	158.81	4.47	2.15	0.11	2.26
	2,	14-16	73.98	159.91	1.84	1.07	0.02	1.09
		43-45	74.26	160.91	1.03	0.30	0.00	0.30
		74-76	74.55	161.94	0.11	0.02	0.00	0.02
		103-105	74.83	162.93	0.02	0.00	0.00	0.00

Hole	Core-section interval (cm)		Depth (cmbsf)	Time (10ka)	Diatom abundance (10 *6/g)	P.sulcata speciments (10 *6/g)	Warm-water speciments (10 *6/g)	Inflow abundance (10 *6/g)
B	9H-2,	132-134	75.11	163.92	0.19	0.00	0.02	0.02
	3,	14-16	75.41	164.99	0.19	0.05	0.00	0.05
		43-45	75.69	165.98	0.09	0.02	0.00	0.02
		74-76	75.99	167.04	0.34	0.13	0.00	0.13
		103-105	76.26	168.00	0.24	0.04	0.02	0.07
		132-134	76.54	169.00	0.02	0.00	0.00	0.00
	4,	14-16	76.85	170.09	0.30	0.64	0.00	0.64
		43-45	77.13	171.09	0.13	0.04	0.00	0.04
		75-77	77.43	172.15	2.59	0.64	0.04	0.69
		103-105	77.70	173.11	1.99	1.01	0.02	1.03
	5,	14-16	78.28	175.17	0.41	0.13	0.00	0.13
			78.43	175.70	Top of Olduvai			
		43-45	78.56	176.24	0.00	0.00	0.00	0.00
		75-77	78.87	177.52	0.02	0.00	0.00	0.00
		103-105	79.14	178.63	0.06	0.02	0.00	0.02
		132-134	79.41	179.75	1.74	0.15	0.04	0.19
	6,	14-16	79.72	181.03	0.17	0.02	0.00	0.02
		43-45	80.00	182.19	0.13	0.07	0.00	0.07
		75-77	80.30	183.43	1.91	0.90	0.00	0.90
		103-105	80.57	184.54	8.57	3.73	0.04	3.77
		132-134	80.85	185.70	8.57	3.69	0.39	4.07
	7,	14-16	81.15	186.94	2.49	0.69	0.04	0.72
		43-45	81.43	188.09	3.39	1.01	0.11	1.12
	10H-1,	39-41	82.29	191.65	0.15	0.06	0.00	0.06
		109-111	82.97	194.46	9.64	2.56	0.10	2.65
	2,	39-41	83.74	197.64	0.39	0.13	0.00	0.13
			83.90	198.30	Base of Olduvai			

Proc. 30ᵗʰ Int'l. Geol. Congr., Vol. 11 , pp. 231-235
Wang Naiwen and J. Remane (Eds)
© VSP 1997

Middle to Late Pleistocene Shallow-Marine Sedimentary and Faunal Cycles Corresponding to the Orbital Precession or Obliquity in the Simosa Group, Boso Peninsula, central Japan

TAKANOBU KAMATAKI[1] and YASUO KONDO[2]

[1] *Division of Earth and Planetary Sciences, Graduate School of Science, Kyoto University, Kyoto 606-01, Japan*
[2] *Department of Geology, Faculty of Science, Kochi University, Kochi 780, Japan*

Abstract

A total of nine depositional sequences corresponding to 20,000 or 40,000 year cycle, were identified in the non-marine to shallow-marine cyclic sequences of the Shimosa Group distributed in the Anesaki district, on the west coast of Boso Peninsula, east of Tokyo Bay. A typical cycle consists of freshwater mud and shallow marine sand (transgressive systems tract), and shallow marine sand (highstand systems tract). Analysis of the molluscan fossil assemblages in the sediments has shown that the depositional environment and the marine climate oscillated between cold-water, fresh- or brackish-water environment or upper sublittoral zone, and warm-water lower sublittoral zone. The water depth change of a scale of about 50-60 m and closely related climatic change strongly suggest a glacio-eustatic origin. Based on published data by fission track dating, the cyclothemic sequence span from the isotopic scale 5 to 11. From the Milankovitch theory, the timing of each cycle is inferred to corresponds to 20,000 or 40,000 yr. cycles caused by the cyclic changes in the orbital precession and obliquity.

Keywords: sedimentary cycle, molluscan fossil, precession, obliquity, the Milankovitch theory, middle to late Pleistocene, Shimosa Group

INTRODUCTION

Sedimentary cycles caused by glacio-eustatic sea-level fluctuation during the Pleistocene have been recognized from onland shallow-marine sedimentary sequences, including Okehu, Kai-iwi, and Shakespeare groups in Wanganui, New Zealand [3, 1], Omma Formation in Kanazawa [9, 10, 11] and Kazusa and Shimosa groups in the Boso Peninsula [2, 24, 12, 25, 6, 4, 13, 18, 14, 5, 7, 8], Japan. Among these, the Boso peninsula is located near the mixed water of the cold Oyashio and the warm Kuroshio (Fig. 1) and the molluscan associations in these sediments are sensitive indicator of climatic changes. Also, paleoclimatic regime shown by the molluscan association can be a proxy for sea-level. Therefore, these sedimentary cycles has been interpreted as originated from glacio-eustatic sea-level fluctuation [*e.g.*, 25, 12, 13, 8]. Recently, Ito and O'Hara [7] claimed that they identified one depositional sequence originated from glacio-eustatic sea-level fluctuation by sequence stratigraphic analysis in the Jizodo Formation, corresponding to the stage 12 to 10 of oxygen isotope stratigraphy. Similarly, Yabu Formation has been correlated to the stage 10 to 8, and Kamiizumi, Kiyokawa, and Yokota formations to the stage 8 to 6 [18, 5, 15]. However, these interpretations have not been verified by quantitative data on molluscan fossil assemblages contained abundantly in the sequences.

Our quantitative analyses on the molluscan fossils have shown that the Jizodo and Yabu Formations, which for have been believed to correspond to 100,000 year cycle, are further

234

subdivided into three and two distinct depositional sequences, respectively. This paper provides a brief summary of water depth and climatic changes through these cycles.

GEOLOGICAL SETTING

Middle to late Pleistocene cyclic sequences of the Shimosa Group is best exposed in the northern part of Boso Peninsula, central Japan (Fig. 1). This Group was deposited in shallow embayment area called Paleo-Tokyo Bay. The Shimosa Group has been subdivided into six formations, that is, the Jizodo, Yabu, Kamiizumi, Kiyokawa, Yokota, and Kioroshi formations, in upward sequence, on the basis of the sedimentary cyclothems in Anesaki district, Chiba Prefecture [24]. These sedimentary cycles are alternating deposits of non-marine mud and marine sand.

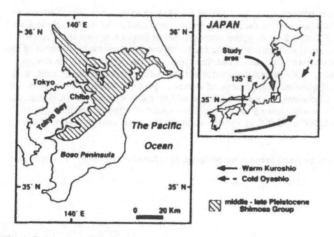

Fig. 1. Index map of the study area.

METHODS

For the quantitative analysis of molluscan fossils, 85 bulk samples of a volume of c. 7,000 cd were collected at each point. Each sample was sieved through a 2 mm mesh screen to separate fossils and sediments. We used the data on recent molluscus, including both bathymetrical and geographical distributions as paleoclimatic indicator [19, 20, 17]. Most of the molluscan fossils from the Shimosa Group are extant species. Oyama [20] defined bathymetrical distribution as follows; N 0: species now living in intertidal zone, N 1: from low tidal mark to 20-30m deep, N 2: from 20-30m to 50-60m, N 3: from 50-60m to 100-120m, and, N 4: from 100-120m to 200-250m. O' Hara [17] used the following distribution of geographical distribution; K-1: species now living in area south of latitude 35° N, K-2: south of lat. 39° N, J: from south of lat. 35° N to north of lat. 39 N, O-2: north of lat. 35° N, and O-1: north of lat. 39° N. We used these definitions in analyzing the fossil assemblages.

SEDIMENTARY FACIES AND MOLLUSCAN FOSSILS

Figure 2 summarizes paleoenvironmental reconstruction for the cyclic sequences of the Shimosa Group. Each depositional sequence of the Shimosa Group, consists of two systems tracts:

transgressive systems tract (TST) and highstand systems tract (HST) (Fig. 2). The TST is represented by estuarine mud, transgressive lag, and offshore muddy sand. The estuarine mud facies formed in an embayment environment in an early stage of rising sea-level. The transgressive lag was deposited in an open-coast environment in an early stage of rising sea-level. The bounding surface between the TST and the HST (downlap surface) can be easily recognized in the Shimosa Group; top of the TST is characterized by weakly cemented densely fossiliferous bed, forming a condensed section. The HST is represented by shoreface and beach sands.

The molluscan fossils preserved within the Shimosa Group are powerful indicators of water-depth and marine climate (Fig. 2). The cyclic change from the upper sublittoral cold-water molluscus, via lower sublittoral warm-water molluscus, again to upper sublittoral cold-water molluscus, appear in each depositional sequence. The base of the TST includes shallow-marine molluscan fossils, such as autochthonous *Crassostrea gigas* and *Potamocorbura amurensis* in estuarine mud or allochthonous *Glycymeris yessoensis* and *Dosinia troscheli* in pebbly sand. The ratio of lower sublittoral warm-water molluscan fossils increase toward the top of the TST. The condensed section includes lower sublittoral warm-water molluscus, such as autochthonous *Keenaea samarangae*, *Glycymeris rotunda*, *Glycymeris pilsbryi* and *Cryptopecten vesiculosus*. While in the HST, upper sublittoral cold-water molluscan fossils, such as allochthonous *Glycymeris yessoensis* and *Pseudocardium sachalinense*, become abundant in upward sequence and no lower sublittoral warm-water molluscan fossils are found. The HST is, therefore, interpreted as formed during an initial stage of sea-level fall under a condition of climatic cooling. These faunal cycles reflect the glacio-eustatic sea-level fluctuation.

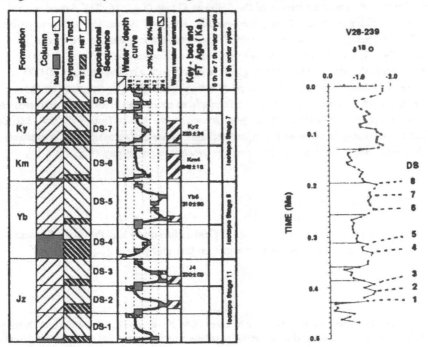

Fig. 2. Summary of the water depth change during the deposition of the Shimosa Group. Oxygen isotope record of V28-239 [23]. Jz: Jizodo Formation, Yb: Yabu Formation, Km: Kamiizumi Formation, Ky: Kiyokawa Formation, Yk: Yokota Formation, DS: depositional sequence, TST: transgressive systems tract, HST: highstand systems tract.

236

DISCUSSION AND CONCLUSIONS

1. The Jizodo Formation is found to comprise three depositional sequences and the Yabu Formation two sequences. Therefore, excluding the Kioroshi Formation located in the uppermost unit of the Group, the Shimosa Group comprises eight depositional sequences, each corresponding to a transgressive-regressive cycle (Fig. 2). Each depositional sequence has characteristic features, such as sedimentary facies succession and relative thickness of TST/HST, although the basic structure of the cycle is common among cycles. For example, the cycle structure is highly asymmetric in the lowermost Jizodo Formation; the TST is less than 3 m, while the HST is much thicker, more than 20 m. In the Kiyokawa Formation, the TST and the HST have almost the same thickness. These differences are interpreted as reflecting different geographical setting within the sedimentary basin. Furthermore, the general trend from asymmetric to symmetric cycle from lower to upper indicates a progressive shallowing of the basin. In addition, each depositional sequence has diffrent types of molluscan association and sedimentary facies within the basal part of the TST, caused by variable sedimentary environment of the early stage of rising sea-level. For example, base of the depositional sequence 1 (Fig. 2), includes autochthonous estuarine molluscus within muddy sediments, while base of the depositional sequence 2 (Fig. 2), includes allochthonous shallow-marine molluscan fossils within transgressive lag deposits.

2. The age of the lower to middle Shimosa Group is c. 0.4-0.2 Ma, based on fission track dating of key-tephras (Fig. 2) and electron spin resonance dating of molluscan shells[eg., 23, 16]. The climatic change and sea-level fluctuation of the order of 100,000 years have been considered to dominate after 0.6 Ma, includes the age of the Shimosa Group [e.g., 26, 21]. Actually, the depositional sequences originated from glacio-eustatic sea-level fluctuation with periods of the order of 100,000 years, were reported from the Pleistocene shallow-marine sediments in the Kanto district [4, 14, 18]. However, each depositional sequence of the Shimosa Group is, interpreted as formed under the influence of glacio-eustacy with a period of the c. 20,000 or c. 40,000-years, which corresponds to the period of orbital precession or obliquity by the Milankovitch theory. The water-depth curve inferred from molluscan fossil association in each depositional sequence is correlated with oxygen isotope curve derived from V28-239 [22]. Each cycle of the lower to middle Shimosa Group corresponds to the transgression and regression in eight cycles of the stage 11, 9, and 7 of the oxygen isotope stratigraphy (Fig. 2).

ACKNOWLEDGEMENTS

We would like to thank Professor Fujio Masuda of Osaka University and Professor Kiyotaka Chinzei of Kyoto University for helpful comments on this study.

REFERENCES

1. S. T. Abbott and R. M. Carter. The sequence architecture of mid-Pleistocene (c. 1.1-0.4 Ma) cyclothems from New Zealand: facies development during a period of orbital control on sea-level cyclicity. Spec. Publs. Int. Ass. Sediment., 19, 367-394 (1994).
2. N. Aoki and K. Baba. 1980, Pleistocene molluscan assemblages of the Boso Peninsula, Central Japan. Sci. Rep., Inst. Geosci., Tsukuba Univ., (Sec. B), 1,107-148 (1980).
3. R. M. Carter, S. T. Abbott, C. S. Fulthorpe, D. W. Haywick, and R. A. Henderson. Application of global sea-level and sequence-stratigraphic models in Southern Hemisphere Neogene strata from New Zealand.

Spec. Publs. Int. Ass. Sediment., 12, 41-65 (1991).

4. M. Ito. High-frequency depositional sequences of the upper part of the Kazusa Group, a middle Pleistocene forearc basin fill in Boso Peninsula, Japan. *Sediment. Geol.*, 76, 155-175 (1992).

5. M. Ito. Stratigraphic records of glacio-eustasy in the Shimosa Group, a middle to late Pleistocene sedimentary fill in Paleo-Tokyo Bay, Japan. *J. Coll. Arts and Sci. Chiba Univ.*, B-26, 31-40 (1993).

6. M. Ito and S. O'Hara. Depositional sequences of the lower part of the Shimosa Group in Kimitsu and Futtsu Cities, Boso Peninsula, Japan. *Rep. Environ. Res. Organization Chiba Univ.*, 16, 1-8 (in Japanese with English abstract) (1990).

7. M. Ito and S. O'Hara. Diachronous evolution of systems tracts in a depositional sequence from the middle Pleistocene palaeo-Tokyo Bay, Japan. *Sedimentology*, 41, 677-697 (1994).

8. T. Kamataki and Y. Kondo. under review (in Japanese with English abstract).

9. A. Kitamura and Y. Kondo. Cyclic changes of sediments and molluscan fossil associations caused by glacio-eustatic sea-level changes during the early Pleistocene - a case study of the middle part of the Omma Formation at the type locality. *J. Geol. Soc. Jap.*, 96, 19-36 (in Japanese with English abstract) (1990).

10. A. Kitamura, H. Sakai, and M. Horii. Sedimentary cycles caused by glacio-eustacy with the 41,000-year orbital obliquity in the middle part of the Omma Formation (1.3-0.9 Ma). *J. Sediment. Soc. Jap.*, 38, 67-72 (in Japanese with English abstract) (1993).

11. A. Kitamura, Y. Kondo, H. Sakai, and M. Horii. Cyclic changes in lithofacies and molluscan content in the early Pleistocene Omma Formation, Central Japan related to the 41,000-year orbital obliquity. *Palaeogeogr., Palaeoclimatol., Palaeoecol.*, 112, 345-361 (1994).

12. Y. Kondo. Faunal condensation in early phases of glacio-eustatic sea-level rise, found in the middle to late Pleistocene Shimosa Group, Boso Peninsula, central Japan. In: *Sedimentary facies in the active plate margin*. A. Taira and F. Masuda (Eds). pp. 197-212, TERRAPUB, Tokyo (1989).

13. Y. Kondo, S. Matsui, and K. Chinzei. Taphonomy and paleoecology of the Pleistocene molluscs in the Boso Peninsula. In: *Island Arcs: Cenozoic stratigraphy and tectonics of Japan*. H. Kato and H. Noro (Eds). pp. 99-108. *Guide Book vol. 2 of 29th IGC Field Trip* B22 (1992).

14. N. Murakoshi and F. Masuda. Estuarine, barrier-island to strand-plain sequence and related ravinement surface developed during the last interglacial in the Paleo-Tokyo Bay, Japan. *Sediment. Geol.*, 80, 167-184 (1992).

15. H. Nakazato. Stratigraphic Relationship between the Kiyokawa and Kamiiwahashi Formations of the Middle-upper Pleistocene Shimosa Group, Chiba Prefecture, Central Japan. *J. Nat. Hist. Mus. Chiba*, 2, 115-124 (in Japanese with English abstract) (1993).

16. H. Nakazato, K. Shimokawa, and N. Imai. ESR age dating of the middle to upper Pleistocene Kazusa and Shimosa Groups. *Earth Monthly (Gekkan Chikyu)*, 12, 37-42 (in Japanese) (1990).

17. S. O'Hara. Molluscan fossils from the Higashiyatsu formation. *Rep. Environ. Res. Organization Chiba Univ.* B-6, 67-83 (in Japanese with English abstract) (1973).

18. H. Okazaki. Sequence stratigraphy of the Shimosa Group. In: *A Plio-Pleistocene fore-arc basin-fill in the Boso Peninsula, central Japan*. Y. Makino, F. Masuda et al. (Eds). pp. 52-60. *Guide Book of 29th IGC Field Trip* A10 (1992).

19. K. Oyama. On the vertical distribution of marine mollusca, *Venus(Japan. J. Malacol.)*, 17, 27-35 (in Japanese with English abstract) (1952).

20. K. Oyama. Revision of Matajiro Yokoyama's Type Mollusca from the Tertiary and Quaternary of the Kanto Area. *Palaeont. Soc. Japan. Spec. Pap.* 17, 148p (1973).

21. W. F. Ruddiman, M. E. Raymo, D. G. Martinson, B. M. Clement, and J. Backman. Pleistocene evolution : Northern hemisphere ice sheets and North Atlantic Ocean. *Paleoceanography*, 4, 353-412 (1989).

22. N. J. Shackleton and N. D. Opdyke. Oxygen-isotope and paleomagnetic stratigraphy of Pacific core V28-239: Late Pliocene to Latest Pleistocene. In: *Investigations of Late Quaternary Paleoceanography and Paleoclimatology*. R. M. Cline and J. D. Hays (Eds). pp. 449-464. *Geol. Soc. Am. Mem.*, 145 (1976).

23. M. Suzuki and S. Sugihara. Fission-track age constraints on the Plio-Pleistocene boundary in the Kazusa Group. *Abs. Ann. Meeting. Japan Assoc. Quat. Res.*, 13, 69-70 (in Japanese) (1983).

24. S. Tokuhashi and H. Endo. Geology of the Anesaki District. Quadrangle Series, scale 1 : 50,000, Geological Survey of Japan, 136 P. (in Japanese with English abstract) (1984).

25. S. Tokuhashi and Y. Kondo. Sedimentary cycles and environments in the middle-late Pleistocene Shimosa Group, Boso Peninsula, central Japan. *J. Geol. Soc. Jap.*, 95, 933-951 (in Japanese with English abstract) (1989).

26. D. G. Williams, R. C. Thunell, E. Tappa, D. Rio, and I. Raffi. Chronology of the Pleistocene oxygen isotope record: 0-1.88 m.y. B.P.. *Palaeogeog., Palaeoclimatol., Palaeoecol.*, 64, 221-240 (1988).